The biology of free-living nematodes

The biology of free-living nematodes

Second edition

By
WARWICK L. NICHOLAS
Reader in the Department of Zoology,
Australian National University

CLARENDON PRESS · OXFORD
1984

Oxford University Press, Walton Street, Oxford OX2 6DP

London Glasgow New York Toronto
Delhi Bombay Calcutta Madras Karachi
Kuala Lumpur Singapore Hong Kong Tokyo
Nairobi Dar es Salaam Cape Town
Melbourne Auckland

and associates in
Beirut Berlin Ibadan Mexico City Nicosia

Oxford is a trade mark of Oxford University Press

Published in the United States
by Oxford University Press, New York

First published 1975
Second edition 1984

British Library Cataloguing in Publication Data

Nicholas, Warwick
The biology of free-living nematodes.—2nd ed.
1. Nematoda
I. Title
595.1'82   QL391.N4
ISBN 0-19-857587-4

Library of Congress Cataloging in Publication Data

Nicholas, Warwick L.
  The biology of free-living nematodes.
  Bibliography: p.
  Includes index.
  I. Nematoda.   I. Title.
QL391.N4N5 1983   595 1'82   83–11437
ISBN 0-19-857587-4

Computerset by Promenade Graphics Limited
Printed by The Thetford Press Limited, Thetford, Norfolk

Preface

MY motive in writing a book was to interest a broad spectrum of biologists in the free-living nematodes, bringing together our knowledge of these animals in a comprehensible and balanced way. I have been encouraged by the interest shown in the first edition by fellow nematologists and biologists of other persuasions to write a second edition. I have retained the same format in this edition, but most of the text has been completely rewritten, though in several short sections, where very little new work has been done, only minor revisions have been made. Several tables, for example, have been reproduced in the second edition where the data seemed worth repeating. Since the first edition was published in 1975, interest in free-living nematodes has greatly increased, with a much wider range of biologists working on free-living nematodes. The enormous growth of publications has made it difficult to keep the lists of references within manageable proportions, and I have reluctantly felt it necessary to omit the titles of individual papers from the bibliography.

It is not my intention to exaggerate the differences between free-living nematodes and parasitic nematodes, and I do not believe any sharp distinction can usefully be made. However, until fairly recently, the parasites have received overwhelmingly more attention from biologists, and I have sought to redress that imbalance. Moreover, by confining my attention largely to free-living nematodes, a relatively smaller book can be produced at less cost and more quickly so that it should not become greatly outdated while in preparation.

A biologist who sets out to write a comprehensive account of a group of animals is unlikely to give an equally satisfactory account of all the various branches of biology involved. It has been put to me that symposia by groups of experts, each covering different fields of biology, are likely to be more useful today than monographs. I agree that when successful, symposia can highlight the expanding fields of interest more effectively, but such symposia, in my experience, seldom provide a comprehensive view of a broad field of biology. Instead, I have aimed at a balanced account of established knowledge, and I must hope that biologists will forgive my many limitations.

The publication of the first edition of my book brought me valuable and rewarding contacts with many nematologists in all parts of the world. Their constructive criticisms and shared interests have amply rewarded me for the efforts I have made, and I hope that this second edition will prove of interest to nematologists and other biologists.

Canberra   W.L.N.
December, 1982

Acknowledgements

I am grateful to Dr Aimorn Stewart for assisting me with the illustrations in several ways. Dr Stewart produced the electron micrographs, both TEM and SEM, reproduced in the book, as well as many others from which I made drawings. She also prepared many specimens for light microscopy and finally labelled the illustrations. I am also very grateful to Mrs Doreen Kjeldsen for typing the manuscript and for her patience with my inconsistencies, errors, and corrections. I appreciate the helpful criticisms of fellow nematologists since the publication of the first edition and hope they will feel that I have made good use of their advice. I would particularly like to thank the following for their advice and comments: Professor Coomans, Dr Geraert, Dr Gerlach, Dr Platt, Dr Riemann, Dr Warwick, Dr Hansen, and Dr Lorenzen. I also thank the many authors who sent me unsolicited copies of their publications. I may perhaps single out Dr Lorenzen, Dr Tsaloliklin, Dr Platonova, and Dr Galtsova for copies of their books, which I should not otherwise have obtained. I am grateful to Roger Pulvers for English translations of Russian.

I thank the following authors for permission to reproduce figures or tables from their work as cited in the text: Dr Albertson, Dr Behme, Professor Coomans, Dr Geraert, Dr Grootaert, Dr Lorenzen, Professor Nigon, Dr Riemann, Dr Sohlenius, Dr Sudhaus, Dr Thomson, and Dr Wyss. I am indebted to a number of publishers for their permission to reproduce the figures and tables listed below:

Figs. 2.4 and 2.25: Annalen van de Koninklijke Belgische Vereniging voor Dierkunde; Figs. 2.5 and 2.15: E. J. Brill, Leiden; Fig. 2.18: Service des Editions de l'ORSTOM, Bondy; Fig. 2.20: The Almqvist & Wiksell Periodical Company, Stockholm; Fig. 2.14 and Table 2.1; The Royal Society, Calton House, London; Fig. 2.19e: Springer-Verlag, Heidelberg; Table 4.2: Gibco Division, The Dexter Corporation, New York; Tables 8.1 and 8.2: Munksgaard International Booksellers and Publishers, Copenhagen.

Contents

1. Introduction

NEMATODES occur everywhere that life can be supported and they are the most numerous of all the metazoa. Nematodes are found in all soils from the polar regions to the tropics, whether arid or humid. They are abundant in the bottom sediments of lakes, rivers, and the oceans. In the oceans they dominate the meiofauna, which comprises the benthic animals too small to be easily seen by the naked eye and too large to be classified as microbes. The biosphere is carpeted with a layer of nematodes several centimetres thick. They are absent from the plankton, except as parasites of other animals, because, although some swim vigorously, buoyancy devices have not generally developed. The only exception I can think of is *Turbatrix aceti*, which swims continuously in vinegar, aided by the buoyancy of lipid reserves. The best known nematodes are those which attack man, his domestic animals, or his agricultural crops, but amongst the great variety of free-living nematodes, one, *Caenorhabditis elegans*, has caught the attention of many biologists with widely differing interests.

The hallmark of nematodes is undulatory propulsion, an efficient means of locomotion in mud, soil, and similar particulate materials. They are essentially aquatic and require at least a film of water in which to be active, but many can tolerate prolonged drying or freezing in a cryptobiotic state, permitting them to inhabit deserts and the polar regions, becoming active when water is present. Because of their small size and the obscurity of their habitats, they are virtually unknown to the layman, except as parasites, and the free-living nematodes have no generally accepted common name in English.

Many attempts have been made to assemble various invertebrate 'pseudocoelomate' groups into a single phylum variously termed the Nemathelminthes or Aschelminthes. Hyman (1951) used the Aschelminthes as a phylum, to include the Nematoda, Gastrotricha, Rotifera, Kinorhyncha, Priapulida, and Nematomorpha as classes in her monographic treatment of the invertebrates, a practice since followed in many textbooks on the invertebrates. The arguments for this classification often rest more on convenience and pragmatism than on phylogeny. Andrassy (1976), in a taxonomic textbook on Nematodes, considers them as a class in the Nemathelminthes, but many other nematologists have been unhappy with the concept of the Aschelminthes as a phylum (Maggenti 1971). Moreover, their common negative character, the lack of a true coelom, seems insufficient grounds to unite the various members of the group, especially since the nature of the nematode body cavity is debatable.

1

In the nematodes, the body cavity, which is fundamental to the hydraulic skeletal systems, appears during embryonic development after gastrulation. It has sometimes been considered a remnant of the blastocoel, or alternatively as a schizocoel arising within the mesoderm as muscles differentiate. However, unlike the true coelom, it lacks an epithelial cell lining (though this has recently been challenged according to Coomans 1979a), and neither the gonads nor the so-called excretory system arise from it. Consequently, its usual designation as a *pseudocoel* may be appropriate.

The Nematoda are morphologically and biologically highly consistent, and clearly and unambiguously separable from the other invertebrates. It seems reasonable to accord the nematodes, as one of the largest groups of invertebrates, the rank of Phylum, instead of pragmatically placing them in the phylum Aschelminthes (as in the previous 1975 edition of my book).

Morphologically the nematodes most closely resemble the Gastrotricha and the Nematomorpha. The structure of the alimentary canal in Gastrotricha is very similar to that of nematodes, and there are also similarities in the cuticle, musculature, and embryology. The gastrotrichs, like the nematodes, are small and aquatic, but unlike the nematodes depend on ciliary locomotion. The Nematomorpha is a very small group of relatively large worm-like parasites of invertebrates, which become free-living when adult. Superficially they resemble nematodes quite closely, and like nematodes propel themselves by undulatory locomotion. The complex cuticle and the longitudinal musculature is also reminiscent of nematodes, and there are more tenuous similarities in their nervous system, gonoducts, pseudocoel, and life-cycle (Hyman 1951).

The only fossil nematodes, from amber and lignite, are relatively recent (Poinar 1977) and do not throw any light on the origin of the Nematoda. The nematodes are probably a very ancient group originating like the other major invertebrate phyla in the Precambrian seas, though a more recent origin from the Arthropoda, on dubious evidence, has had its proponents (Rauther 1909 quoted by Riemann 1977a). The most characteristic morphological features of the Nematoda, are associated with undulatory propulsion and one may speculate that the adoption of this system of locomotion allowed the penetration of the bottom sediments of the ocean, opening up a vast habitat which nematodes dominate today. With undulatory locomotion, muscles replace cilia as the motive organs (Riemann 1977a). Animals from many phyla inhabit the interstices of marine sediments, the meiofauna, and tend to be relatively small, elongated, and composed of few cells (Swedmark 1964). The nematode body is composed of relatively few cells, typically 1000 non-gonadial cells, in predetermined locations arising by a fixed sequence of cell divisions. Asexual reproduction, developmental regulation, and regeneration have not developed. Parasitic nematodes may become much larger, but then have larger cells, while retaining relatively few cells and rigidly determined development. These aspects of nematode developmental biology strongly suggest that the ancestral nematodes were inhabitants of the meiofauna. However, the method of undulatory propulsion, dependent on a cuticle and a hydraulic skeleton, may itself have imposed a strictly determinate pattern of development and a high degree of morphological uniformity on nematodes.

A paper by Harris and Croften (1957) on *Ascaris suum*, a gastrointestinal parasite of pigs, showed how its cuticular structure was related to the high internal pressure of its body fluids, and was a major conceptual advance. *Ascaris* is much larger than most nematodes, greatly facilitating measurements of internal fluid pressures, but making it atypical. However, Croften and Harris's conclusions explained some of the characteristic morphological features of nematodes, showing the relationship between cuticle structure, body fluid pressure, and undulatory propulsion which are applicable to most nematodes.

In *Ascaris* contraction of the longitudinal muscles of the body wall, together with the very stiff elasticity in the cuticle maintains a very high internal hydrostatic pressure. Localized contraction of either the dorsal or ventral muscles will tend to shorten the body and increase the cross-sectional area, while retaining a constant volume. The cuticle resists any increase in cross-sectional area and the body flexes with localized stretching of the opposing muscle block. In *Ascaris* the cuticle contains three layers of essentially inextensile collagenous fibres, arranged in a spiral basket-work around the body, set at an angle of about 75° to the longitudinal axis of the body. Such a system gives a stiff, flexible body, ideally suited to the undulatory locomotion of a smooth elongated body through a viscous medium. Harris and Croften (1957) suggested that many of the characteristic features of nematodes are a consequence of this mechanical system, a conclusion which has been generally supported by subsequent work. Free-living nematodes have complex cuticles, apparently designed to bend but to resist radial expansion, though differing in construction from that in *Ascaris*. In some a system of interlocking plates may functionally replace the need for high internal fluid pressure (Riemann 1977a), while a few aberrant nematodes have different methods of locomotion with corresponding morphological peculiarities. The high internal hydrostatic pressure has been used to explain many other aspects of morphology, for example, the structure of the alimentary canal and the gonads, the lack of functional cilia, and the lack of a recognizable osmoregulatory organ (Harris and Croften 1957).

Parasitic nematodes have received much more attention than free-living nematodes, but the two cannot be easily separated. Parasitism has clearly arisen independently many times within the Nematoda and takes many different forms. It is an ecological and physiological relationship between the nematode and its host, which cuts across taxonomic categories. For the practical purpose of writing this book, the parasites of vertebrates can be set aside on taxonomic grounds without much difficulty, though many have free-living stages in their life-cycle. With other parasites it is more difficult.

The order Tylenchida includes plant- and fungal-feeding nematodes as well as endoparasites of insects and other invertebrates, and a few even parasitize vertebrates (Siddiqi 1980). It includes most of the species which attack higher plants and many nematologists would classify all of these as either ecto- or endoparasites (Norton 1978). However, it may be stretching this classification too far to include those which range freely through the soil feeding on plant roots or on fungal mycelia, though both habits may occur within one genus or even species. Most zoologists would consider parasitism to imply a sustained physical and physiological dependence of an animal on its host. It is clearly convenient for the plant pathologists to consider all the plant-pathogenic species (including many Dorylaimida as well as Tylenchida) as plant parasites. For my part, in writing this book, I shall consider the fungal-feeding species, but largely ignore the nematodes dependent on higher plants, as well as nematode–higher plant relationships. These have been the subject of many other textbooks. I shall draw on material from the Tylenchida when it best illustrates some general aspect of nematode biology.

The Tylenchida have until recently been considered divisible into two suborders, the Tylenchina and the Aphelenchina, both characterized by a similar mouth stylet which can puncture plant cell walls. A very comprehensive review by Siddiqi (1980) has led him to separate the two taxa as orders, the Tylenchida and Aphelenchida, because he believes they have separate phylogenetic origins. I have accepted this view, though it may lead to some ambiguities in discussing the Tylenchida because this was not the previous practice. The Aphelenchida includes plant-feeding ectoparasites, fungal-feeding species, insect and annelid parasites, and predators on other nematodes (Siddiqi 1980).

The Dorylaimida, a more diverse taxon, contains several families with mouth stylets which can puncture plant cells, though of a different type to that found in the previous two orders. Some are very important to plant pathologists, damaging agricultural plants directly or by transmitting pathogenic viruses, but most move freely through the soil, and do not show such an intimate association with their plant food as is found in many families of Tylenchida.

The Rhabditida are predominantly free-living, most feeding on bacteria, and will contribute much to this book, but they also include some parasites and various borderline cases. There are species in which a parasitic generation alternates with a free-living generation (Rhabdiasidae) and others which facultatively or of necessity enter the bodies of invertebrates to complete their life-cycles. Some Rhabditida, which can complete their life-cycle by feeding on micro-organisms in the soil, may enter the bodies of potential invertebrate hosts, where they remain quiescent until the host dies, whereupon they feed on the putrefying cadaver. In other species the nematode precipitates the death of its 'host' by the introduction of pathogenic bacteria, then feeds on the putrefying carcass. Poinar (1972) has discussed many facultative parasites of insects.

The Nematoda are divided into two classes, the Adenophorea and the Secernentea. The Secernentea, which includes the Rhabditida, Tylenchida, and Aphelenchida, are primarily terrestrial and freshwater, and includes most of the plant and animal parasites. The Adenophorea are predominantly marine and contain relatively few parasites of animals or plants, with the exception of one large taxon, the Mermithoidea, all of which are parasitic in invertebrates. Almost all the known species of mermithids are terrestrial, though one deep-sea species *Rhaptothyreus typicus* is known (Hope 1977). Surprisingly, most parasites of marine mammals, birds, reptiles, and even fish, belong to the Secernentea, though there are few marine free-living Secernentea, and very few feed on plants. One genus of Tylenchida, *Halenchus*, is marine. Perhaps all the Secernentea, and the Dorylaimida and Mermithoidea from the Adenophorea, have evolved together with the terrestrial higher plants, insects, and vertebrates, displacing more primitive nematode taxa (Maggenti 1971; Ferris, Goseco, and Ferris 1976). The Adenophorea may have evolved in the sea very

TABLE 1.1

The classification of nematodes which have been extensively used for research

	Family	Natural food and habitat
Class Secernentea		
Order Rhabditida		
Caenorhabditis elegans	Rhabditidae	Bacteria, soil
Caenorhabditis briggsae	Rhabditidae	Bacteria, soil
Turbatrix aceti	Cephalobidae	Bacteria, vinegar
Panagrellus redivivus	Cephalobidae	Bacteria, fermenting foods
Panagrellus silusiae	Cephalobidae	Bacteria, fermenting foods
Order Tylenchida		
Ditylenchus dipsaci	Tylenchidae	Plant cells, endoparasite
Meloidogyne javanica	Heteroderidae	Plant cells, endoparasite
Order Aphelenchida*		
Aphelenchus avenae	Aphelenchidae	Fungal mycelia, soil

* The order Aphelenchida has only recently been raised to ordinal rank by Siddiqi (1980), and was previously considered a suborder Aphelenchina within the Tylenchida.

much earlier than the Secernentea with relatively few now inhabiting freshwater and terrestrial environments. Marine invertebrates are less commonly parasitized by nematodes than terrestrial invertebrates (Osche 1966 quoted by Sudhaus 1974*a*). The paucity of parasitic marine Adenophorea remains unexplained.

Most of the experimental work on free-living nematodes has concentrated on only a few species. I have listed their taxonomic position and full names in Table 1.1. There has been an extraordinary growth of interest in one species of free-living nematode, *Caenorhabditis elegans*, in recent years. Many of those working on this species would not consider themselves nematologists, but primarily geneticists, developmental biologists, molecular biologists, cell biologists, or belonging to some other branch of biology. They have become interested in this species because of the advantages it possesses as an experimental animal for many fields of biology. Their interest was stimulated initially by Dr S. Brenner, who selected this species at Cambridge, UK in 1965 for work on the genetic basis for the development of the nervous system (Riddle 1978).

Important advances by Brenner and his numerous co-workers have come from the use of genetic mutants. Cultures of mutant lines can be preserved in a liquid N_2 refrigerator, and, to facilitate work with mutants, a *Caenorhabditis* Genetics Center has been established in the USA (Division of Biological Sciences, 106 Tucker Hall, University of Missouri, Columbia, Mo. 65211, USA). Interest in this species is now world-wide with many laboratories actively working on the species, and it has been the subject of international meetings. A *Caenorhabditis* newsletter has been established (edited in 1981 by Dr R. S. Edgar, Thinmann Labs., University of California, Santa Cruz, Ca. 95064, USA).

Brenner was influenced in his choice of *C. elegans* by the late E. D. Dougherty, who had pioneered axenic cultures of nematodes, and from whom cultures were obtained (Riddle 1978). The isolate used by Brenner, *C. elegans* Bristol, was collected by me from mushroom compost (provided by Dr Staniland from near Bristol, UK) in 1956 and put into axenic culture by myself (Nicholas, Dougherty, and Hansen 1959).

Caenorhabditis briggsae, a related species from California, had earlier been cultured by Dougherty, and both species were maintained for several years in the same laboratory in California. Many studies of *C. briggsae* have been reported, mostly on nutrition and biochemistry. It has since turned out that there may have been an accidental mix-up in the stock cultures, and genetic analysis has shown that cultures maintained in some laboratories as *C. briggsae* are *C. elegans* (Friedman, Platzer, and Eby 1977). The two are so similar that it is no easy matter to tell them apart, though they are good species.

I shall be discussing much work on both species in this book and have decided not to attempt to rename the species retrospectively. It should be borne in mind that some of the papers published on *C. briggsae* may in fact describe work on *C. elegans*. Vanfleteren (1980) has listed some laboratories which may have reported work under an incorrect name. Another isolate of *C. elegans* from France was collected by Nigon, *C. elegans* Bergerac, and has been widely used, especially by French workers.

Most biologists working on *C. elegans* have chosen to work with a nematode because of the advantages this species has as an experimental animal for work on problems of general biological significance. Advances in our knowledge of nematodes are a secondary consideration. Their interests are succinctly summed up in the title of a two-volume symposium entitled 'Nematodes as biological models' (edited by Zuckerman 1980) which concentrates very largely on *C. elegans*. Research is providing new insights into the genetic control of development and into such problems as ageing. I shall not enter into a discussion of nematodes as models of biological phenomena. My aim, so far as work on *C. elegans* is concerned, is to ask what it tells us about the biology of nematodes. It may well be, as is persuasively argued by Zuckerman in the symposium, that research on ageing in *C. elegans* will throw light on ageing in man. Whether this turns out to be true or not, it certainly tells us much about ageing in nematodes.

Much of the most valuable work has involved genetic analysis of *C. elegans* and the manipulation of mutant alleles. I have thought it best to leave the genetic mapping of this species and description of the many mutant alleles aside. It seems better to leave such details to those actually engaged in the research, to whom the details are most clearly relevant.

Nematology is well served with specialized scientific journals and I have listed the best known of these in Table 1.2.

TABLE 1.2
Journals publishing papers in nematology

Nematologica	E. J. Brill, Leiden, The Netherlands
Journal of Nematology	The Society of Nematologists, USA
Revue de Nématologie	Office de la Recherche Scientifique et Technique Outre-Mer, France
Proceedings of the Helminthological Society of Washington	Washington, USA
Journal of Helminthology	London School of Tropical Medicine, London, UK
Helminthologia	Academia Scientiarum, Slovaca
Indian Journal of Helminthology	Helminthological Society of India
Indian Journal of Nematology	The Nematological Society of India
Helminthological Abstracts, Series B, Plant Nematology	Commonwealth Bureau of Helminthology, Herts, UK

2. Morphology

2.1. General plan, shape, and size

Free-living nematodes are small, usually less than 2.5 mm long when fully grown, and may not exceed 0.1 mm in length, as in *Trichoderma*. The Dorylaimida includes the largest terrestrial nematodes, some reaching 6–7 mm in length, while Antarctic marine nematodes 5 cm long are known (Platt 1978). Taking published data on the size of adult female nematodes, a log normal distribution will fit the size frequency distribution of free-living species (Kirchner, Anderson, and Ingham 1980).

The body is an elongated cylinder, tapered to varying degrees at both ends, appearing threadlike to spindle-shaped, depending on the ratio of length to maximum width. This ratio (*a* in De Man's formula, commonly used in taxonomic descriptions) ranges from 10 to 140 or more (Filipjev 1921). The elongated body is ideally suited to the undulatory locomotion characteristic of nematodes.

The wet weight of small nematodes is generally calculated from their volume, measured by microscopy (Andrassy 1956; Warwick and Price 1979) and their specific gravity, determined by flotation in liquids of different density. Specific gravities from 1.084 to 1.13 have been reported (Overgaard Nielsen 1949; Bair 1955; Andrassy 1956; Wieser 1960). The figures come from marine, freshwater, and terrestrial nematodes.

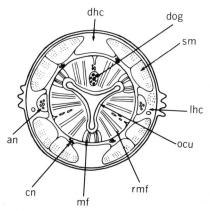

FIG. 2.1. Diagrammatic transverse section through the oesophageal region of a nematode. dhc, dorsal hypodermal chord; dog, duct of dorsal oesophageal gland; sm, somatic longitudinal muscles; lhc, lateral hypodermal chord; ocu, oesophageal cuticular lining; rmf, radial muscle fibrils; mf, marginal fibrils; cn, cephalic sensory nerve; an, amphidial nerve.

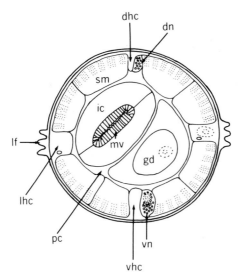

FIG. 2.2. Diagrammatic transverse section through the intestinal region of a nematode. dhc, dorsal hypodermal chord; dn, dorsal motor nerve; vn, ventral motor nerve; pc, pseudocoelom; lhc, lateral hypodermal chord; lf, lateral field; ic, intestinal cell; mv, microvilli; gd, gonad.

Andrassy (1956) and Yeates (1973*a*) have listed the estimated weights of many species. The average weight of 50 non-marine species was 1.628 μg, ranging from 0.025 to 15.001 μg (Andrassy 1956). *Dorylaimus stagnalis* was 600 times as heavy as *Monhystera simplex*. The dry weight is generally 20–5 per cent wet weight (Wieser 1960; Sivapalan and Jenkins 1966; Buecher and Hansen 1971).

A cuticle encloses the body which consists essentially of two concentric tubes, separated by the body cavity or pseudocoelom. The outer tube, Figs. 2.1 and 2.2, consists of the cuticle, the epidermis (or hypodermis) which gives rise to the cuticle, and the longitudinal musculature, while the inner is formed by the more or less straight alimentary canal, consisting of the buccal cavity (or stoma), the oesophagus (or pharynx), the intestine, and the rectum. The oesophagus and rectum, but not the intestine, are lined by cuticle. The reproductive organs, which are tubular, lie in the pseudocoelom, the male usually opening with the anus to form a cloaca, the female usually by a separate opening, the vulva. Since most nematodes are transparent, the internal organs can be seen in 'whole mounts' under the microscope. With well 'fixed' and prepared specimens, the cells and cell nuclei can be seen, and even the ducts and openings of unicellular glands. These are often better seen in anaesthetized specimens with Normarski differential interference contrast optical microscopy (Sulston and Horvitz 1977).

Three different genera of terrestrial free-living nematodes are illustrated in Figs. 2.3, 2.4, and 4.1. These are respectively species of *Rhabditis*, bacteria-feeding soil nematodes; *Aporcelaimellus*, algal-feeding soil nematodes; and *Mononchus*, predatory soil nematodes.

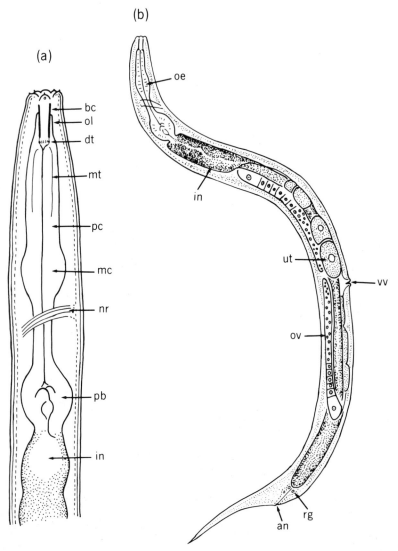

FIG. 2.3.　Female *Rhabditis*. (a) cephalic region; (b) whole body. oe, oesophagus; in, intestine; ut, uterus containing eggs; vv, vulva; ov, ovary, rg, rectal glands; an, anus; bc, buccal cavity; ol, oesophageal sleave; dt, denticles in buccal cavity; mt, marginal tubes; pc, procorpus; mc, metacorpus; nr, nerve ring; pb, posterior bulb or post corpus.

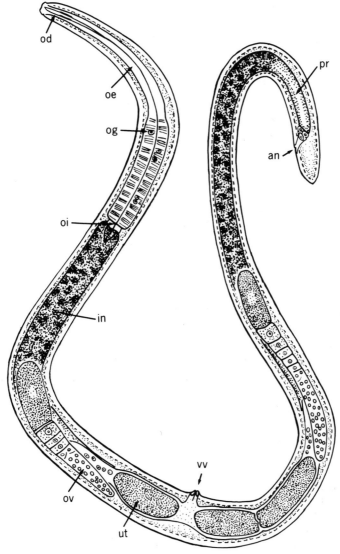

Fig. 2.4. *Aporcelaimellus* female. od, odontostyle; oe, oesophagus; og, nucleus of oesophageal gland; oi, oesophago-intestinal valve or cardia; in, intestine; ov, ovary; ut, uterus containing an egg; vv, vulva; pr, posterior intestine or pre-rectum; an, anus.

2.2. Symmetry and nomenclature

The nematode body shows two fundamental symmetries: a bilaterally symmetrical body, upon which is superimposed the triradial symmetry of

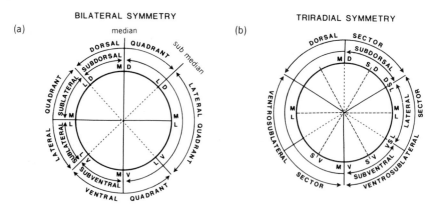

FIG. 2.5. Nomenclature based on symmetry: (a) bilateral; (b) triradial. MD, mediodorsal; LD, laterodorsal; ML, mediolateral; LV, lateroventral; MV, medioventral; SD, subdorsal, DSL, dorsosublateral; SV, subventral; VSL, ventrosublateral. (Reproduced from Coomans (1978), *Annls. Soc. r. Zool. Belg.* **108**, 115–18).

the head. Triradial symmetry is shown by the cephalic nerves, the sensilla, the mouth, the lips, and the anterior region of the alimentary canal. The existence of two different patterns of symmetry has led to nomenclatorial problems, but an acceptable, if rather complicated terminology has been worked out by Coomans (1978), following a discussion by nematologists at the 'First workshop on the systematics of marine free-living nematodes' in Ghent in August 1977. This terminology, which is a further development of that used by De Coninck (1965), is illustrated in Figs. 2.5(a), (b), and it is to be hoped that it will be followed in the future by all nematologists.

When referring to bilateral symmetry, there are four equal quadrants, dorsal, ventral, and two lateral separated by laterodorsal and lateroventral radii, and subdivided by mediodorsal, medioventral, and mediolateral radii. The regions limited by these radii should be identified by the prefix *sub*. In triradial symmetry there are three equal areas, one dorsal and two ventrosublateral, called sectors, to distinguish them from the quadrants, and limited by two dorsosublateral and one medioventral radii respectively. The dorsal sector, larger than the dorsal quadrant, is subdivided into two subdorsal sectors. The ventrosublateral sectors are divided into lateral and subventral sectors. Confusion also arises in describing the radially arranged cells of the oesophagus and Grootaert and Coomans (1980) suggest uniform terminology (Fig. 2.6).

2.3. The cuticle

The cuticle is a highly complex structure, in which the electron microscope shows systems of fibres, plates, or struts, embedded within several layers of

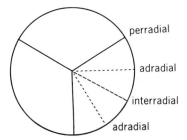

FIG. 2.6.   Nomenclature for the oesophageal radial symmetry. (Reproduced from Grootaert and Coomans (1980), *Nematologica* **26**, 406–31.)

protein, of varying consistency, of which collagen is a major constituent. Within it lie the ducts of hypodermal glands and the dendrites of peripheral nerves. In free-living nematodes it is usually about 2 µm thick and moulted four times during the life of the nematode. Cuticle also lines the buccal cavity, oesophagus, and rectum.

In many nematodes the cuticle is annulated, and often forms longitudinal ridges, folds (alae), or grooves, and there may be setae, especially in marine nematodes. The surface features are most clearly seen by scanning electron microscopy (SEM), as in Plates 1–3, and good examples of the use of SEM have been published by Sher and Bell (1975) and De Grisse and

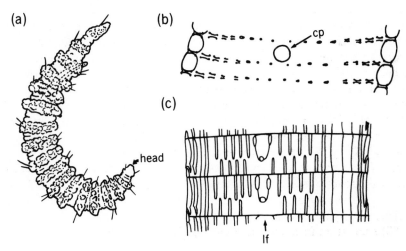

FIG. 2.7.   Example of cuticular structure in three marine nematodes as seen by optical microscopy. (a) *Desmoscolex* (Desmoscolecidae) showing very prominent annulation; (b) *Pomponema* (Cyatholaimidae) showing cuticular punctation and a pore; (c) *Actinonema* (Chromadoridae) with punctation, annualation and lateral fields. cp, cuticular pore; lf, lateral field.

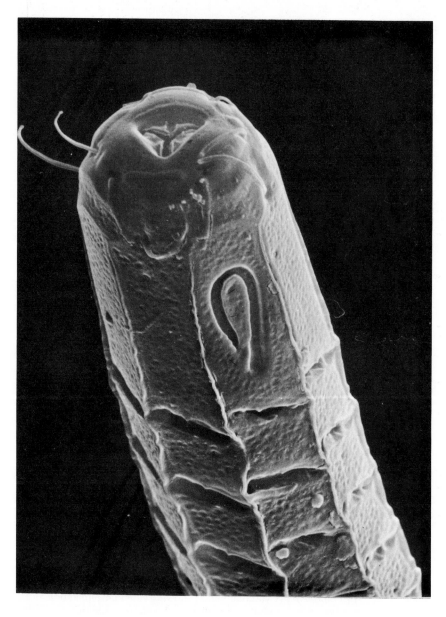

PLATE.1. Head of *Cyttaronema* (Ceramonematidae).

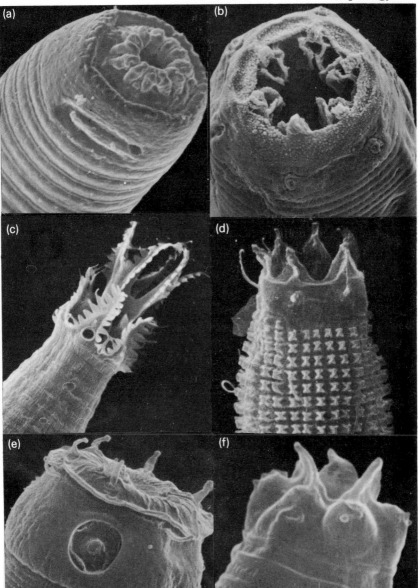

PLATE 2. The heads of six free-living nematodes. (a) *Actinonema* (Chromadoridae) showing amphid and buccal cavity; (b) *Halichoanolaimus* sp. showing six
labial papillae; (c) *Acrobeles* (Cephalobidae) showing furcate labial probolae,
alternating with non-furcate cephalic probolae, and circular amphid; (d) *Gonionchus* (Xyalidae) showing labial flaps and prominent cuticular ornamentation; (e)
Desmodora cazca (Desmodoridae) with circular amphid and labial papillae; (f)
Acrobeloides (Cephalobidae) showing labial probolae, labial papillae and small
amphidial opening on lateral lip.

PLATE 3. Cuticular structure and accessory male sexual organs. (a) Supplementary organs of *Onyx* (Desmodoridae); (b) lateral fields of *Acrobeloides* (Cephalobidae); (c) spicules, sensory papillae, and lateral field of *Desmodora cazca* (Desmodoridae); (d) spicules and sensory papillae on *Acrobeles* (Cephalobidae); (e) cuticular ornamentation and lateral field of *Stegelletea* (Cephalobidae); (f) cuticular plates of *Cyttaronema* (Ceramonematidae).

Lagasse (1969). Indeed, SEM photomicrographs are becoming normal parts of taxonomic descriptions.

Sometimes cuticular annulations are too fine to be resolved by light microscopy, but in others coarse annular rings are very obvious (see Fig. 2.7(a)). Wallace (1970) shows by calculation that annulation increases the flexibility of the cuticle, without buckling, because less force is required in undulatory locomotion to deform annuli, than to stretch or compress a smooth cuticle. He suggests that annulations are formed in the developing cuticle by buckling of the cortical layers as the tightly coiled larval nematode moves within the egg shell.

The cuticle overlying the lateral hypodermal chords often differs in fine structure from that elsewhere, as in many Tylenchida (Johnson, Van Gundy, and Thomson 1970a), and frequently forms a number of ridges which run the whole length of the body (shown in Plate 4). In Chromadorida there is often a break in the pattern of cuticular punctation (see Fig. 2.7(c)) with a groove marking the lateral field, or a lateral keel, which may aid in locomotion (Riemann 1976).

Bird (1971) proposed that the cuticle is basically a three-layered structure, with the outer subdivided into two; a proposal which has been generally accepted by other nematologists, though sometimes with differing terminology. Examples of equivalent terminology are:

Bird (1971)		Johnson and Graham (1976)	De Grisse (1977)
Cortical {	external	Epicuticle	Epicuticle
	internal	Outer median	Exocuticle
Median		Median	Mesocuticle
Basal		Basal	Endocuticle

A different interpretation of cuticular fine structure with differing terminology has been given by Maggenti (1979), to which I will return after discussing Bird's interpretation.

On its outer surface the external cortical layer is bounded by a trilaminar layer, 3–10 nm thick, with outer and inner thin osmiophylic zones. Bonner, Menefee, and Etages (1970), from their study of the parasitic nematode *Nematospiroides dubius*, interpreted this trilaminar membrane as a cell membrane (plasmalemma), and suggested that the cuticle was an intracellular product of the hypodermal cells. Bird (1980) in reviewing the structure of the epicuticle concludes that it shows many similarities to a cell membrane from which it probably evolved.

Negatively charged carbohydrates coat the outer surface of the cuticle of *Caenorhabditis briggsae* (Himmelhoch, Kisiel, and Zuckerman 1977; Himmelhoch and Zuckerman 1978). The affinity of ^{125}I-labelled plant lectins for this 'glycocalyx' in *C. briggsae* and *C. elegans* shows that it

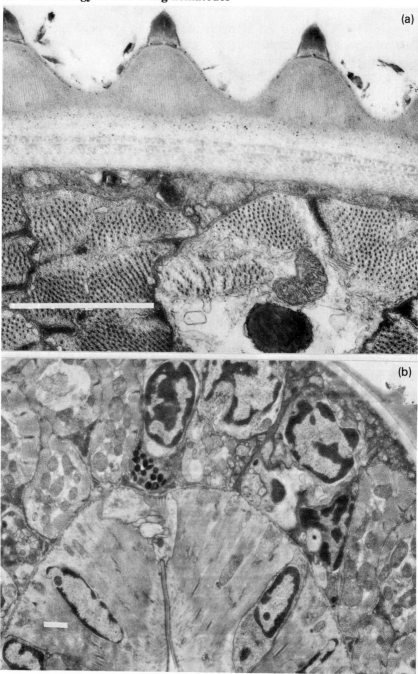

PLATE 4. Electron micrograph of transverse section through: (a) cuticle and body-wall muscles of *Omicronema* (Xyalidae); (b) oesophagus. (Bar = 1μm.)

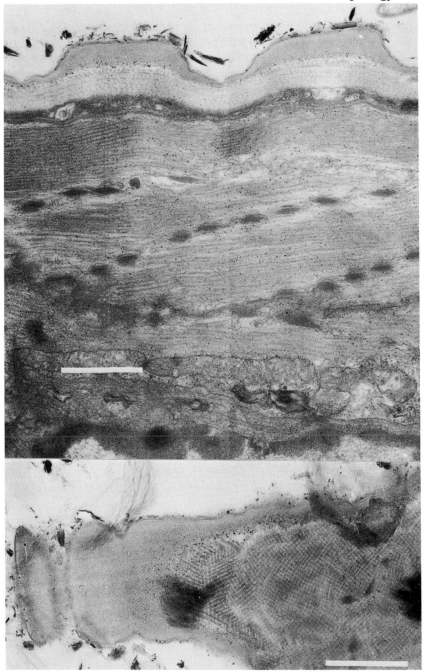

PLATE 5. Electron micrograph of longitudinal section through cuticle and body-wall muscles of *Omicronema*. (Bar = 1 μm.)

contains galactose, glucose, mannose, and N-acetylglucosamine (Zuckerman, Kahane, and Himmelhoch 1979). However, since enzymic digestion did not show these hexoses to be present as neuraminic, hyaluronic, nor glucuronic acids (Himmelhoch and Zuckerman 1978), it differs from the typical glycocalyx of mammalian cell membranes, as well as from that of several animal parasitic helminths (Lumsden 1975). The cuticle of *Meloidogyne javanica* may also possess a negatively charged surface, but again, apparently not due to neuraminic acid (Himmelhoch, Orion, and Zuckerman 1979).

The external and internal cortical layers are distinguished by differences in their density and fine structure. The median layer is usually less dense than the cortical layers and may be semi-fluid, but often contains supporting struts, rods, or plates. The basal layer, or the lower median layer, often contains fibres or short rods. The fine structure of the cuticle of *Omicronema* (Xyalidae) is shown in Plates 4 and 5 and diagrammatically in Fig. 2.8. Supporting struts set in a fluid median layer are found in *C. briggsae* (Zuckerman, Himmelhoch, and Kisiel 1973*a*). In Chromadoridae

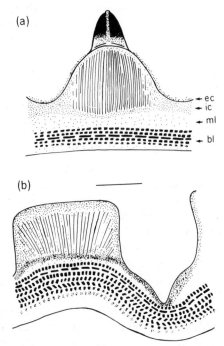

Fig. 2.8.   Diagrammatic representation of cuticular structure in *Omicronema* (Xyalidae) as revealed by electron microscopy. (a) Transverse section; (b) longitudinal section. ec, external cortical layer; ic, internal cortical layer; ml, median layer; bl, basal layer. (Bar = 1 μm.)

(a)   (b)

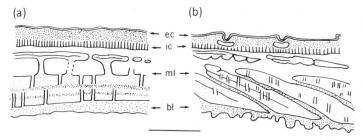

FIG. 2.9. Diagrammatic representation of cuticular structure in *Dichromadora* (Chromadoridae) from electron microscopy. (a) Transverse section; (b) longitudinal section. ec, external cortical layer; ic, internal cortical layer; ml, median layer; bl, basal layer. (Bar = 1 μm.)

the median layer contains overlapping plates, like tiles on a roof, penetrated by radial canals, as in *Euchromadora* (Watson 1965) and *Dichromadora* (see Fig. 2.9).

In *Acanthonchus duplicatus* (Cyatholaimidae) closely spaced tubular rods within the median layer support the cortical layer (Wright and Hope 1968). When the rods are viewed end on with the light microscope, through the superficial layers of the cortex, they appear as dense arrays of refractile points—or punctations (Fig. 2.7(b), (c)). Such punctations are common in Chromadorida.

Pores are found in the cuticle of many marine nematodes and in Cyatholaimidae there appear two types, which Sharma, Hopper, and Webster (1978) termed type I (or hypodermal), and type II (or lateral modified punctations), both numerous in *A. duplicatus*. The fine structure of type I has been described (Wright and Hope 1968). The pore opens to the exterior by a transverse slit, set in a pit leading into a canal supported by a dense ring, which traverses the cuticle to a hypodermal cell. Two sublateral rows of hypodermal glands opening by cuticular pores are present in *Chromadorina germanica* (Chromadoridae) (Lippens 1974).

In their classical study of the hydrostatic skeleton and locomotion of *Ascaris*, Harris and Croften (1957) emphasized the importance of the three layers of circumferential collagenous fibres, which form a basket-work within the basal layer. Circumferential fibre layers are found in other large nematodes such as Dorylaimida, though not necessarily in the basal layer. Such an arrangement of cuticular fibres in the large terrestrial genus *Aporcelaimellus* is illustrated in Fig. 2.10. Some marine nematodes, notably Desmodorida, have strongly annulated cuticles. In the Ceramonematidae, the cuticle is composed of closely interlocking plates (Haspeslagh 1979), as can be seen in Plates 1 and 2. These cuticles may be forming an articulated external skeleton, replacing the hydrostatic skeleton, but with some loss of flexibility.

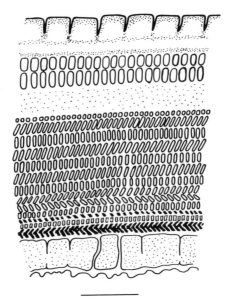

Fig. 2.10.   Diagrammatic representation of an electron micrograph of a longitudinal section through the cuticle of *Aporcelaimellus* sp. (Dorylaimida) showing
many layers of electron lucent fibrils, cut at varying angles to their long axes, and
set in a more electron dense matrix. (Bar = 1 μm.)

In small terrestrial nematodes a striated basal layer is common, probably
giving structural support. Samoiloff (1973) has suggested a role in cuticular
transport for the striae (p. 23). A striated basal layer is found in
Tylenchida (Johnson *et al.* 1970*a*), and the larvae but not adults of *C.
briggsae* (Epstein, Castillo, Himmelhoch, and Zuckerman 1971; Cassada
and Russell 1975). Wisse and Daems (1968) concluded that the striated
layer in the larval *Globodera* (=*Heterodera*) *rostochiensis* (Tylenchida)
was formed of 'vertical' rods 20 nm apart, bound by thin filaments, in a less
dense matrix. Popham and Webster (1978) interpret the striations
differently in *C. elegans*, as sections of intersecting membranes. They see
the layer consisting of osmiophobic protein blocks separated by osmiophylic lipoprotein membranes, the latter longitudinal and circumferential.
     Fine structural details of the cuticle have been described in Rhabditida:
Panagrellus silusiae (Samoiloff and Pasternak 1968, 1969); Tylenchida:
Pratylenchus penetrans (Kisiel, Himmelhoch, and Zuckerman 1972),
Hemicycliophora arenaria, *Hirschmanniella gracilis*, *Hi. belli*, Aphelenchida: *Aphelenchus avenae* (Johnson *et al.* 1970*a*), several species of
Meloidogyne (Bird and Rogers 1965; Bird 1968; Johnson and Graham
1976), and *Heterodera* (Shepherd, Clark, and Dart 1972). In the
plant-parasitic genus *Heterodera*, the larvae and males have typical

tylenchid cuticles, but the female forms a dormant resistant cyst. In some species one, and in others two, thick additional collagenous layers are laid down within the cuticle, and these are tanned by polyphenols (Shepherd *et al*. 1972).

In *Deontostoma californicum* (Enoplida), as described by Siddiqi and Viglierchio (1977), the thick cuticle, 10–13 nm thick, has nine distinct layers. The inner cortical layer is radially striated and is penetrated by wide and narrow canals from the exterior. The median layers possess radial rods and the basal layer longitudinal fibres.

Maggenti (1979) believes that attempts to relate the fine structure of nematode cuticles to three layers, cortex, median, and basal, leads to drawing false homologies. Instead he believes that there are basically four layers, often subdivided into sublayers, but that not all four are present in all taxa. He believes that the complete series is most consistently present in the Enoplia, and takes *Deontostoma* (Enoplida) as representative. He recognizes: a thin *Epicuticle*, typically trilaminar; an *Exocuticle*, usually subdivided and often with radial striae; a *Mesocuticle*, composed of rods, fibres, plates, or struts in a less dense ground substance, often in organized sublayers, and an *Endocuticle*, poorly defined and often with extensions from the hypodermis. Tylenchida possess only the epi- and exocuticles as do larval and some adult Rhabditida.

2.4. Moulting (=ecdysis)

The shedding of the old cuticle during moulting has been described by light microscopy in many nematodes, but electron microscopy is required to follow much of the detail. The cuticle over the external body surface is a product of the hypodermis, but in the oesophagus, and possibly the rectum, it is formed by musculoepithelial cells. The cuticular lining of the stomodaeum, including its stylets and valves, and of the rectum are shed, as are the ducts of all hypodermal glands. Moulting is generally preceded by a period of inactivity, or lethargus (see Section 6.4).

In *Panagrellus silusiae* (Rhabditida) the old cuticle does not separate until a new cuticle has formed beneath it from material synthesized within cells of the hypodermal chords. Samoiloff and Pasternak (1969) found no evidence of digestion and reabsorption of the old cuticle. The transfer of proteins, labelled with [^{3}H-] leucine, from the hypodermal chords to the cuticle through invaginations in the basement membrane has been demonstrated autoradiographically by Samoiloff (1973), who suggests the striated layer may be a transfer system within the cuticle.

In *Aporcellaimus sp.* (Dorylaimida), during the first moult, the new cuticle similarly forms beneath the old cuticle from material synthesized in the hypodermal cells, which increase in depth prior to moulting (Grootaert and Lippens 1974). The structure of the desmosomes between the cells of the hypodermal chords changes before moulting from zonulae adherentes to septate desmosomes, which may permit moulting secretions to act on the cuticle. The old cuticle then separates. No evidence has been found of reabsorption. As in *P. silusiae*, differentiation of the cuticular fine structure is completed after separation of the old cuticle. In *Meloidogyne javanica* (Tylenchida), the products of the digested inner layers may be reabsorbed through the new cuticle (Bird and Rogers 1965). Moulting in several other Tylenchida has been described in fine-structural detail by Johnson, Van Gundy, and Thomson (1970*b*).

Bonner *et al.* (1970), who studied the moulting in the animal parasite *Nematospiroides dubius*, suggest that the new cuticle forms by intracellular secretion, transforming the outer layers of the hypodermal cells into cuticle, which remains bounded by a plasmalemma.

Moulting has been described by light microscopy in many Rhabditida (Maupas 1899; Jantunen 1964; Thomas 1965; Hechler 1967; Yarwood and Hansen 1969; Chin and Taylor 1970); Tylenchida (Hechler and Taylor 1966*b*; Roman and Hirschmann 1969); and Dorylaimida (Lamberti 1969). There are minor differences as to how the nematode escapes from the ruptured old cuticle and which stomodeal parts are passed through the gut.

2.5. Hypodermis (=epidermis); glands

The hypodermis encloses the animal in a single cell layer beneath the cuticle, to which it gives rise. It is thickened dorsally, laterally, and ventrally to form the hypodermal chords. The lateral chords are more prominent than the dorsal and ventral. The dendrites and axons of the nervous system lie within the hypodermal cells. In small nematodes the interchordal cytoplasm may be very thin with the nucleated regions of all the hypodermal cells within the chords. In *C. elegans* the hypodermis consists of several syncytial tissues together with non-syncytial *seam* cells. The seam cells, lying in the lateral chords, divide during post-embryonic development to give rise to a number of structures, including the alae along the lateral fields (Singh and Sulston 1978). Mutants lacking seam cells and their derivatives can develop to the adult stage.

The hypodermis contains unicellular glands associated with the cephalic sense organs, which will be discussed later (see Section 2.11). Caudal glands are present in most Adenophorea, excluding most Dorylaimida, but are not found in Secernentea (see Fig. 2.23). Their mucoid secretions, extruded through a spinneret at the tip of the tail, enable aquatic nematodes to attach themselves temporarily to the substratum. There are usually three gland cells, with prominent nuclei and nucleoli, lying in the pseudocoel behind the anus, though in some taxa the cells lie anterior to the anus. The secretion passes down extensions of the cells to the spinneret, which can be closed by a central plug. Lippens (1974) has described the fine structure of caudal glands of *Chromadorina germanica* (Chromadorida).

Secernentea lack caudal glands, but instead possess paired lateral phasmids in the caudal region, though they are probably sensory receptors and not functionally equivalent to caudal glands. Typically they consist of a gland cell, and associated sensory neurone, which opens through a cuticular canal at a minute pore. In some Tylenchida, the phasmids are located near the middle of the body, dorsal to the lateral fields (Siddiqi 1978). Other hypodermal glands associated with sensory neurones occur along the body of *Acanthonchus duplicatus* (Chromadorida) (Lippens 1974b). Each merocrine unicellular gland cell opens to the exterior by a cuticular pore. Each gland cell is closely associated with a glial cell and a sensory neurone, from which two dendrites (modified cilia) project into the duct.

2.6. Musculature

A single layer of obliquely striated muscle cells lies beneath the hypodermis, separating it from the pseudocoelom, except where four hypodermal chords divide the muscles into four blocks, two dorsal and two ventral. The contractile elements do not occupy the whole cell, but are restricted to parts of the cross-section and descriptive names have been applied to their differing arrangement within the cells, e.g. coelomyarian and platymyarian (Hope 1969). The fine structure of coelomyarian obliquely striated muscles in *Omicronema* is illustrated in Plates 4 and 5, and diagrammatically in Fig. 2.11. Most work has been done on the muscles of *Ascaris suum*, because its large size facilitates classical electrophysiological work (De Bell 1965; Rosenbluth 1965, 1967, 1969; Jarman 1976; Johnson and Stretton 1980), but the three-dimensional structure of the muscles of the large marine Enoplid *Deontostoma californicum* has also been described (Hope 1969). However, it is with *C. elegans* that a combination of electron microscopical, biochemical, biophysical, and genetic methods has made great advances in our knowledge. The use of many mutant lines, with defective muscle cells and impaired mobility, has opened a new dimension in understanding the developmental biology of muscles (Zengel and Epstein 1980).

In *C. elegans* the disposition of all the muscles and their neuromuscular connections are known in great detail (White, Southgate, Thomson, and Brenner 1976). There are 95 muscle cells, 24 in each quadrant, arranged diagonally in two rows. In nematodes, muscle cell processes extend to the motor neurones to form synapses, rather than the axons of the motor

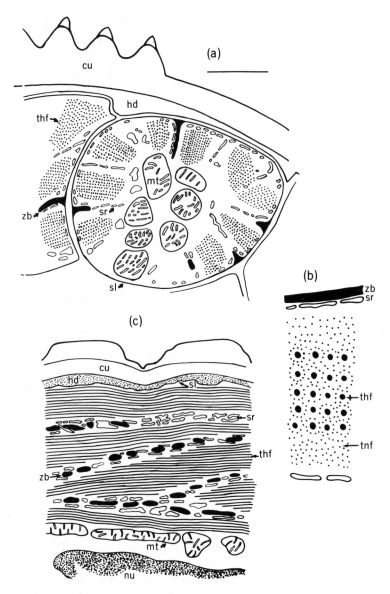

Fig. 2.11. The fine structure of the obliquely striated body-wall muscles of *Omicronema* (Xyalidae) as seen by electron microscopy: (a) transverse section (Bar = 1 µm); (b) an enlargement showing arrangement of thick and thin filaments, the latter omitted from (a) and (c). The precise number of thin filaments surrounding each thick filament has not been determined precisely; (c) a longitudinal section (with thin filaments omitted). cu, cuticle; hd, hypodermis; thf, thick filament; zb, z-body; sr, sarcoplasmic reticulum; mt, mitochrondria; sl, sarcolemma; tnf, thin filaments; nu, nucleus.

neurones finding the muscle cells as in other animals. Extensions from the anterior muscle cells of *C. elegans* pass the nerve ring, turn, and form synapses on the inner surface of the nerve ring. Extensions from the posterior muscle cells reach the dorsal or ventral nerve chords, where they interdigitate, forming synapses with one another and motor neurones. Extensions from intermediate muscle cells form synapses on both the nerve ring and along the nerve chords.

The basic elements of the contractile apparatus of cross-striated muscles, as described in vertebrate striated muscles, are also present in obliquely striated muscles, but differently arranged. The A-bands, I-bands, and H zones are recognizable electron microscopically, formed by the arrangement of the thick filaments and interdigitating thin filaments (see Fig. 2.11 and Plates 4 and 5), but these are differently arranged in the sarcomeres to those in cross-striated muscles. In place of the Z-lines, at 90° to the long axis of the sarcomere, there are rows of dense bodies which form the boundaries between the sarcomeres. While the myofilaments are parallel to the body's long axis, the sarcomeres are not, and both the thick and thin filaments are in a staggered arrangement across the cell, giving rise to oblique striations.

Obliquely striated muscles must have physiological advantages over cross-striated muscles for undulatory propulsion. Alexander (1979) has suggested what these advantages might be by analogy with comparisons between molluscan obliquely striated muscle and vertebrate cross-striated muscle. The longer filaments in obliquely striated muscle make it possible to exert greater tension per unit cross-sectional area, and at lower energy cost, but the maximal rate of contraction is correspondingly slower. The greater overlap of thick and thin filaments means that tension can be exerted over a relatively greater proportional change in length.

The proteins of the contractile apparatus have been extracted from *C. elegans* body-wall muscles and purified (Harris and Epstein 1977; Harris, Tso, and Epstein 1977). The biophysical and biochemical properties of myosin, paramyosin, actin, tropomyosin, and their reactions with ATP, Ca and Mg ions have been demonstrated (Zengel and Epstein 1980). In these properties the proteins from *C. elegans* closely resemble those of other animals. As in other invertebrates, myosin surrounding a core of paramyosin forms the thick filaments. ATPase- and actin-binding sites are located on the heads of the myosin molecules. Actin is the major component of the thin filaments and associated with tropomyosin.

There are two different myosins present in each body-wall muscle cell of *C. elegans*, which are synthesized in fixed proportions during post-mitotic growth of the cells, and which associate as homodimers (Schachat, Harris, and Epstein 1977; Schachat, Garcea, and Epstein 1978; Garcea, Schachat, and Epstein 1978; Mackenzie, Schachat, and Epstein 1978; Mackenzie,

Garcea, Zengel, and Epstein 1978). Mutant alleles which restrict the synthesis of one of these myosins, or paramyosin, disorganize the growth of sarcomeres, leading to progressive paralysis. A third myosin species occurs in the oesophagus. Oesophageal muscles are differently organized and are discussed in Section 2.8. Small muscles, such as dilators, are usually single sarcomeres, and typical cross-striated muscles may occur in *Trichodorus porosus* (Bird 1970).

2.7 The head, lips, mouth, and buccal cavity (=stoma)

The head generally tapers to a terminal mouth without a sharply defined head. In some nematodes a strengthening of the anterior cuticle, or lack of annulation, marks a distinct head region. Its most significant features are three sets of sensilla, but these will be discussed later together with the nervous system (see Section 2.11). The mouth is almost always centrally placed at the tip of the head, leading to a distinctive chamber, variously called the buccal cavity, buccal capsule, or stoma, between the mouth and oesophagus. Rarely, the mouth is offset, dorsally in the Diplopeltidae, and in a few nematodes the buccal cavity is virtually absent, so that the mouth opens into the oesophagus, as in *Halalaimus*. The mouth and buccal cavity are triradiate, and usually surrounded by three or six lips, often taking the form of simple protruberances set off from the smooth contour of the head to varying degrees. They may take a more elaborate form, with deep incisions between each lip. In some nematodes they support thin tenuous flaps, as in some Enoplidae and Xyalidae. These seem characteristic of nematodes found in marine sands and perhaps function as filters (Riemann, personal communication).

Very complex labial appendages are found in some terrestrial Cephalobidae (Sauer, Chapman, and Brzeski 1979). Flap-like extensions of the cuticle, probolae, with feathered margins (tines), may extend from the head in front of the mouth, with an outer ring of single or double probolae, alternating with an inner ring of larger labial probolae (see Plate 1). One of the labial probolae is dorsal; two are ventrosublateral. Their function in these common bacteria-feeding soil nematodes, usually associated with decomposing plant material, is unknown.

The buccal cavity or stoma shows great diversity in form, reflecting the food of the nematodes. It is often divisible into regions, supported by strengthened cuticular plates, and may possess mouth parts variously described as teeth, denticles, stylets, jaws, or mandibles. Examples of different kinds of buccal cavity are drawn in Figs. 2.12, 8.2, and 8.3. Many attempts have been made to identify homologies for its structures in different nematode taxa and to use these to reconstruct phylogenies. The most generally favoured approach has been to seek homologies with the

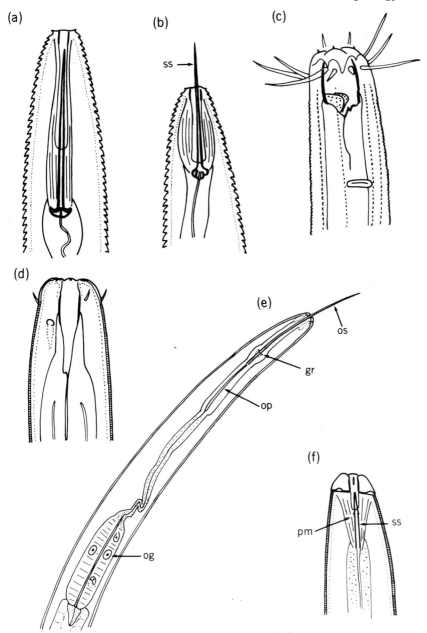

FIG. 2.12. Examples of the buccal cavity in terrestrial nematode genera. (a) *Criconemoides*, stomatostyle withdrawn; (b) *Criconemoides*, stomatostyle protruded; (c) *Prismatolaimus*; (d) *Plectus*; (e) *Longidorus*; (f) *Aphelenchus*, stomatostyle withdrawn. ss, stomatostyle; os, odontostyle; op, odontophore; gr, guide ring; og, oesophageal gland nucleus; pm, protractor muscle.

cuticular subunits of the Rhabditid buccal cavity, originally named by Steiner (Weingartner 1955; Goodey 1963; Gerlach 1966; Siddiqi 1980). Since, however, the Rhabditidae are no longer considered the most primitive family, the rationale for taking this as the basic form has greatly weakened. Naming structures according to their presumed phylogenetic derivation leads to difficulties, when our knowledge of the phylogenetic history of nematodes is so imprecise and controversial, and I accept Cooman's advice that at present descriptive terms without phylogenetic implications are to be preferred. However, the term cheilostome, derived from Steiner's terminology, can often usefully be applied to the anterior buccal cavity, within the lips, with strong cuticular supports for the specialized buccal musculature.

Wright (1976) has drawn attention to the difference between the cuticle of the body surface and the structures of stomodaeal origin and suggested that the region of the alimentary canal modified for the initial incorporation of food be termed the buccal capsule, while the term stoma, which has been used as a synonym for the buccal capsule (Chitwood and Chitwood 1950), should be restricted to structures of stomodeal origin. Some nematodes would then have astomatous buccal capsules. During embryonic development, the anterior alimentary canal is formed by two separate intrusions of ectodermal origin. The first and largest, usually described as the stomodaeum, gives rise to the oesophagus, while the second and smaller has been thought to form the buccal cavity. The two can often be distinguished subsequently by differences in the fine structures of their cuticular lining, but more comparative studies are needed before the strict association between development and fine structure can be assured, and their juxtaposition in the buccal cavity in many taxa is known. In the meantime, it seems best to avoid terms which imply an ontogenetic or phylogenetic derivation and refer to the pre-oesophageal (= pre-pharyngeal region) as the buccal cavity.

A very detailed study has been made by Inglis (1964) of the head in the order Enoplida, tracing several lines of development in the buccal cavity and associated musculature, which reach their greatest complexity in the Enoplidae. Some Enoplids, which feed on fine particulate food, have simple buccal cavities, e.g. *Anticoma*; but in many, the buccal cavity is capacious, permitting the ingestion of large particles for mastication before food is passed to the oesophagus. In the Phanodermatidae, Enoplidae, and Leptosomatidae, fusion of the anterior oesophagus with the body wall over a region with strengthened cuticle, forms a cephalic helmet. In some the anterior end of the oesophagus has also been reinforced to form the oesophageal capsule. Anterior to the oesophagus, a fluid-filled cavity, the cephalic ventricle, occupies the space between the body wall and the buccal cavity, which is reinforced by cuticular bars, also providing support for sensory nerves. In the Enoplidae the three cuticular plates, reinforcing the buccal cavity, may develop into jaws or may support a tooth (*onchium*) (see Fig. 2.13(a), (b)). In other families different parts of the buccal cavity are developed. In Oncholaimidae, for example, the buccal cavity is enlarged at the expense of the anterior oesophageal musculature (Fig. 2.13(c)). In Ironidae the teeth sweep forward and outward, pivoting on the cuticular lining of the oesophagus (van der Heiden 1975).

Tylenchida and Aphelenchida possess similar pointed hollow mouth stylets, which can be used to penetrate plant cells, fungal mycelia, or the bodies of other animals including nematodes. In plant and fungal feeders the stylet is used like a hypodermic syringe to inject the secretion of the

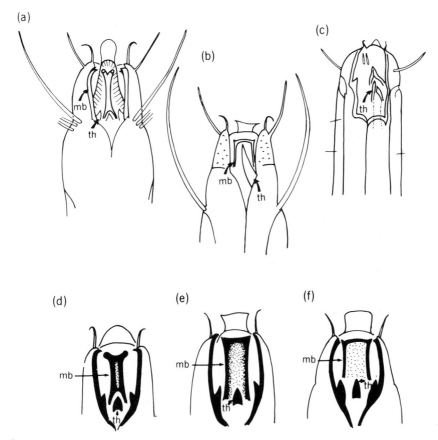

Fig. 2.13.   The buccal cavity and development of teeth and mandibles in representative genera of Enoplidae. (a) *Enoploides*; (b) *Oxyonchus*, showing only one of two ventrosublateral teeth; (c) *Oncholaimus*; (d) *Epacanthion*; (e) *Mesacanthion*; (f) *Enoplolaimus*; mb, mandible; th, tooth.

oesophageal glands into the cells and then suck out the contents. It can very rapidly and repeatedly be protruded from the mouth and withdrawn. The stylet is known as a stomatostyle (or onchiostyle), because of its presumed derivation from the buccal cavity or stoma. It generally has a conical tip, cylindrical shaft, a lumen about 0.2 μm in diameter, and three basal knobs (one dorsal, two ventrosublateral) (see Fig. 2.12). The stylet lies within a conical vestibule or more flexible cuticle extending inward from the mouth. The stylet is replaced at each moult (Roman and Hirschmann 1969). Three, six, or more protractor muscles, run from the anterior body wall, the head framework, or anterior vestibule to the stylet knobs.

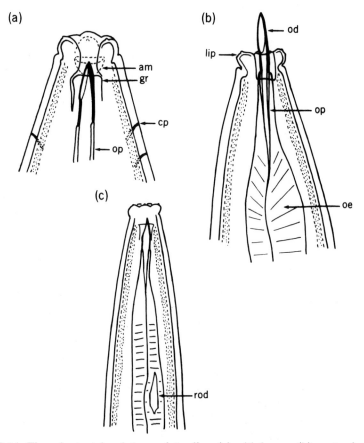

FIG. 2.14. The odontostyle of *Aporcelaimellus*. (a) withdrawn; (b) protruded; (c) replacement odontostyle in larva before final moult. am, amphid; gr, guide ring; cp, cuticular pore; op, odontophore; lip, lip; od, odontostyle; oe, oesophagus; rod, replacement odontostyle.

The fine structure of the stomatostyle has been described in several Tylenchida, e.g. *Ditylenchus dipsaci* (Yuen 1967), *Pratylenchus penetrans* (Chen and Wen 1972), *Tylenchorhynchus dubius* (Byers and Anderson 1972), *Heterodera glycines* and *Meloidogyne incognita* (Baldwin and Hirschmann 1976), and *Hexatylus viviparus* (Shepherd and Clark 1976*b*). Seymour (1975*b*) applied photoelastic stress analysis to perspex models of the stylet of *Heterodera cruciferae*. The shape of the stylet maximizes the ratios of tensile strength to weight, and stiffness to weight. Its modulus of elasticity suggests it may be formed from collagen. The stomatostyle of Aphelenchida, though similar, has probably had a different evolutionary history (Siddiqi 1980). The fine structure of the stylet in *Aphelenchoides blastophthorus* (Shepherd, Clark, and Hooper 1980) shows interesting differences, especially the sensory neurone associated with its shaft (see Section 2.11).

The Dorylaimida displays a variety of powerfully armed buccal cavities. The Monochidae are predators with capacious buccal cavities armed with powerful teeth (Grootaert and Wyss 1979) (see Fig. 4.1). Some families have a hollow buccal stylet which can be used to puncture plant cells or feed on other animals, in a similar way to Tylenchida, but formed differently (see Fig. 2.14). The dorylaimid stylet, which develops from oesophageal tissue is termed an odontostyle. It probably evolved from a ventrosublateral oesophageal tooth, like that found in predatory Nygolaimidae (Coomans 1964).

A thin folded cuticle (sheath) joins the base of the odontostyle to the wall of the buccal cavity at the guide ring. A semi-fluid material lies beneath the cuticle sheath, termed hydrostatic substance (Taylor and Robertson 1971; Grootaert and Wyss 1978), permitting the protrusion of the odontostyle from the mouth in feeding, involving protractor, retractor, and dilatore buccae muscles (Coomans 1964). The fine structure of the feeding apparatus has been described in *Aporcelaimellus* and *Labronema* by Lippens, Coomans, De Grisse, and Lagasse (1975); Grootaert and Wyss (1978), and Coomans and Grootaert (1980). The odontostyle is shed at each moult, and a replacement odontostyle forms early in each intermoult within a ventrosublateral cell in the anterior oesophagus, just behind the functional stylet. It is secreted into a vacuole, its shape determined by the shape of the vacuole with its cytoskeleton of fibrils. Later it moves posteriorly and hardens, only to move forward again to replace the old stylet at moulting (Coomans and De Coninck 1963; Coomans and van der Heiden 1971; Grootaert and Coomans 1980).

Longidoridae transmit plant viruses (Lamberti, Taylor, and Seinhorst 1975) which has stimulated studies of their fine structure (Wright 1965; Lopéz-Abella, Jiménez-Millán, and Garcia-Hidalgo 1967; Aboul-Eid 1969*b*; Taylor, Thomas, Robertson, and Roberts 1970). They possess a very long thin odontostyle, with the supporting oesophageal tissue, the odontophore (or spear extension), better developed than in other Dorylaimida (see Fig. 2.12). The oesophageal lumen is connected to the odontostyle lumen through the odontophore. The odontostyle of *Xiphinema index* has a longitudinal split which Roggen (1975) argues will cause it to twist, facilitating penetration. Bending forces will develop as the long slender stylet is forced into plant tissue, which will induce shearing stresses in the cross-section, which will be asymmetrically distributed because of the split leading to torque.

The Trichodoridae possess yet another type of stylet, formed from the dorsal oesophageal tissue (Raski, Jones, and Roggen 1969). In *Trichodorus christiei* the anterior oesophagus is muscular and supports an outer partially hollow spear, into the lumen of which is inserted a finer inner spear (Hirumi, Chen, Lee, and Maramorosch 1968). Both spears arise from the dorsal wall of the oesophagus, the inner inserted into the outer by way of a groove, and both are hollow for part of their length with solid tips. Food from punctured cells flows past the spear through the continuous lumens of the buccal cavity and oesophagus.

2.8 Oesophagus (or pharynx)

The oesophagus is a muscular tube linking the buccal cavity with the intestine, which when it expands draws food in through the mouth and when it relaxes drives food into the intestine with a pumping action. Some nematologists believe it would be more appropriately called a pharynx, but since both terms are derived from totally different mammalian structures, I shall use oesophagus as the more commonly applied name. The oesophagus, which develops from the stomodaeum, includes muscular, fibrous, glandular, and nervous tissue, and secretes a cuticular lining. A diagrammatic cross-section with triradiate lumen is shown in Fig. 2.1 and terminology in Fig. 2.6.

In its simplest form it is a uniform cylinder, ending in an oesophago/intestinal valve, or Cardia, but often shows one, two, or rarely more, muscular expansions called oesophageal bulbs, which convert the oesophagus into a multi-stage pump. Valves are necessary to prevent regurgitation of food, and are often formed from a cuticular elaboration of the lining of one of the bulbs. These may also act mechanically on the food as it passes through. In many nematodes, the lumen expands at the tips of the oesophageal radii to form three longitudinal canals, visible in whole mounts of many nematodes for varying distances along the oesophagus, the *marginal tubes*. The form of oesophagus reflects the food of the nematode and is often characteristic of the family. Some examples are illustrated in Fig. 2.15.

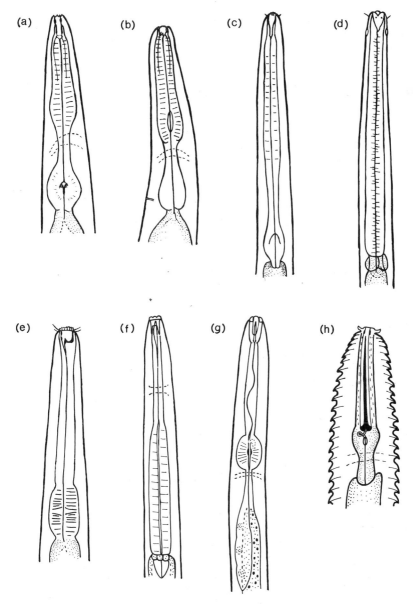

FIG. 2.15. Examples of oesophageal structure. (a) *Rhabditis* (Rhabditida); (b) *Micoletzkia* (Rhabditida); (c) *Plectus* (Leptolaimina); (d) *Monhystera* (Monhysterida); (e) *Ptycholaimellus* (Chromadorida); (f) *Aporcelaimellus* (Dorylaimida); (g) *Aphelenchus* (Aphelenchida); (h) *Criconemoides* (Tylenchida). (a), (b), and (c) are terrestrial bacteria-feeding nematodes; (d) from freshwater, probably feeds on aggregates of bacteria; (e) marine, probably feeds on algae; (f) terrestrial, feeds on algae; (g) on fungal mycelia; and (h) on plant roots.

Uninucleate glands within the oesophagus open by ducts into the lumen. Typically there are said to be one dorsal and two ventrosublateral glands, but in Dorylaimida there are often five uninucleate glands (Loof and Coomans 1970), and this may turn out to be the usual number (see below). The nerves found in the oesophagus presumably co-ordinate its operation and may also control glandular secretion.

A theoretical analysis of the cylindrical soft-walled oesophagus has been given by Roggen (1973). The lumen is opened by the contraction of regularly spaced radial myofibrils, and closed by hydrostatic pressure when they relax. Geometric analysis of its changing configurations during

TABLE 2.1

Cells in the oesophagus of Caenorhabditis elegans *from Albertson and Thomson (1976)*‡

Cell type	Number of cells	Number of nuclei	Cell type	Number of cells	Number of nuclei
Motor neurones			*Muscle cells*		
M1	1	1	m1	1	6
M2	2	2	m2	3	6‡
M3*	2	2	m3	3	6
M4	1	1	m4	3	6
M5	1	1	m5	3	6
			m6	3	3
Interneurones			m7	3	3
I1†	2	2	m8	1	1
I2†	2	2			
I3†	1	1	Total		37
I4	1	1	*Marginal cells*		
I5**	1	1	mc1	3	3
I6†	1	1	mc2	3	3
			mc3	3	3
Other neurones					
neurosecretory-motor, NSM*	2	2	Total		9
motor-interneurone, MI	1	1	*Epithelia cells*		
marginal cell neurones, MC*	2	2	e1	3	3
			e2	3	3
Total		20	e3	3	3
			Total		9
			Gland cells		
			g1	2	3
			g2	2	2
			Total		5

Total cells in pharynx (=oesophagus) = 80‡

* Neurones with mechanoreceptive-like ending attached to muscles with restricted cytoplasm.
** Neurone attaches to motor neurones M2 cell bodies with a junction to *.
† Neurones with mechanoreceptive-like ending.
‡ Corrected as requested by Albertson (personal communication).

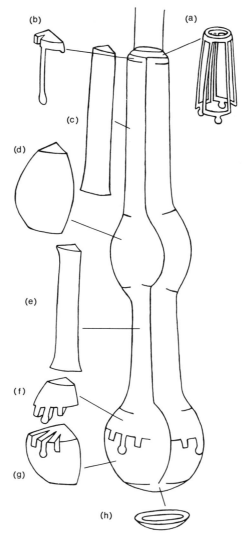

FIG. 2.16. Muscle cells in the oesophagus (= pharynx) of *Caenorhabditis elegans*. The distribution of the eight muscle layers are indicated, with a single cell pictured for each layer. (a) Muscle layer m1 is a single cell which completely surrounds the triangular lumen. Six thin processes run posteriorly for 40 μm, adjacent to the outside edges of the marginal cells. These cross to the middle of each sector to join a binucleate cell body in the nerve cords; (b) posterior to m1, the marginal cells appear so that M2, although forming a thin sheet around the lumen, similar to m1, is divided into three cells. Each projection of M2 in the nerve cords terminates in a single nucleus. Muscle layers M3 (c), m4 (d) and m5 (e) are composed of three wedge-shaped cells, lying one to a sector, as delineated by the marginal cells. Each cell contains two nuclei, one at either side of the nerve cord. Muscles m4 are more spherical than the others, forming the well-developed metacorpus. In the terminal bulb three T-shaped cells (f), slot into the seventh set of three cells (g). A saucer-shaped cell (h), lines the posterior wall of the oesophagus (= pharynx). (Reproduced from Albertson and Thomson (1976). *Phil. Trans R. Soc. Lond.* **B275**, 299–325.)

opening and closing, shows that a triradiate lumen will minimize stresses in the bordering membranes, and tension in the myofibrils, while providing structural stability. The hydraulics of a muscular bulb was considered in a subsequent paper (Roggen 1979). A spherical muscular bulb develops less suction, but greater pressure, than a comparable cylindrical muscular tube of equivalent length and lumen volume. Placing the bulb posteriorly, next to the intestine, will increase its effectiveness in forcing food into the intestine against the hydrostatic pressure of the pseudocoelom (see also Section 2.13).

The structure of the oesophagus of *C. elegans* is known in great detail. The disposition and ultrastructure of the 80 cells forming the oesophagus has been reconstructed by electron microscopy of serial sections by Albertson and Thomas (1976), who have also mapped neuronal connections in great detail. There are 34 muscle cells, nine epithelial cells, five gland cells, and 20 neurones (Table 2.1 and Fig. 2.16), but many of the cells are evidently multi-functional. A basement membrane completely encloses the oesophagus.

The nine epithelial cells form the anterior region of the procorpus, arranged in three successive tiers of three cells, with those of each tier overlapping two cells of the succeeding tier. Their nuclei lie in a long narrow posterior extension from the cells. Posterior to the epithelial cells, three tiers of three marginal cells form the oesophagus, one cell in each tier at the apex of the triradiate lumen (perradial), thereby separating the dorsal from the two ventrosublateral muscle sectors. Three cords (interradial) run within each muscle sector from the anterior end of the oesophagus to the posterior bulb containing neuronal processes, epithelial cell nuclei, some muscle-cell nuclei, and, in the dorsal cord, the duct of the dorsal gland. Intracellular filaments cross the oesophagus from the cuticle to the basement membrane (where there are half-desmosomes). In the metacorpus (first bulb), neuronal processes from the three cords join forming an oesophageal nerve ring; in the posterior bulb a commissure joins the ventrosublateral and dorsal cords.

Two types of glands, g1 and g2, occupy much of the posterior bulb. There are three g1 cells, two anterior ventrosublateral with ducts which open at the anterior of the isthmus, and one posterior dorsal g1 with its duct opening just behind the buccal cavity. There are two posterior latrosubventral g2 cells in the posterior bulb, with ducts opening into the valve region.

There are eight tiers of muscle cells, with one to three cells per tier, each of which typically contains one sarcomere, with its radially orientated myofilaments running from the cuticle to the basement membranes, attached by half-desmosomes. In the posterior bulb the arrangement is more complex, with longitudinal and radial myofilaments associated with the valves. One muscle cell, tier 8, enclosed by the oesophageal basement membrane, is surrounded by five cells of the oesophageal intestinal valve (which has a cuticular lining), which are not included in the oesophageal cell numbers. Some muscle cells have fused to give binucleate cells.

Twenty cells make up the nervous system, and these have been drawn individually by Albertson and Thomson (1976), with their connections shown diagrammatically. They identified seven motor neurones, six interneurones, and five other neurones. Most oesophageal neurones are unbranched, unipolar or bipolar cells, and form synapses *en passant*, instead of typical synaptic endings. The nervous system shows a high degree of bilateral symmetry, with either paired neurone cell bodies in the ventrosublateral sector, or a single unpaired cell body in the dorsal sector forming paired branches. The motor neurone cell bodies either lie in the anterior bulb, associated with the oesophageal nerve ring, or in the posterior bulb. The paired branches of the dorsal sector motor neurones innervate each ventrosublateral sector, with one exception, returning to innervate the dorsal sector. Interneurones may similarly be either paired and ventrosublateral, innervating both a ventrosublateral and the dorsal sector, or dorsal with more complex bilateral branching. Paired bipolar cells, within the anterior bulb, which appear from their fine structure to be both neurosecretory and motor, innervate the isthmus muscles. A single unipolar motor/interneurone forms muscle and neuronal connections around the nerve ring. A pair of bipolar cells innervate the marginal cells in the anterior isthmus.

The oesophagus appears subject to both intrinsic and extrinsic control. A single pair of interneurones (no. 11) receives synapses from the central nerve ring, by way of connections through the basement membrane and ventrosublateral chords in the first muscle tier. Several neurones have their endings bound to adjacent cells by desmosomes, just below the oesophageal cuticle. These are probably mechanoreceptors, while other probable mechanoreceptive endings are located on muscle cells and gland cells. From detailed analysis of the neuronal circuitry, the authors propose a model of the neuronal control of the oesophagus. Inhibition of oesophageal pumping could come from the somatic nervous system through its input to interneurones no. 11. Pumping cycles would be generated intrinsically, responding to mechanoreceptors in the oesophagus and controlled by oesophageal interneurones.

The fact that the triradiate oesophagus is controlled by a basically bilaterally symmetrical nervous system is important for understanding the evolution of nematodes. The presence of five oesophageal gland nuclei, as in the Dorylaimida, rather than three, as generally recognized in nematodes, is interesting. Of more general zoological interest is the multifunctional nature of the cells. Some neurones are both interneurones and motor neurones, while both motor and neurosecretory activity apparently occurs in two other neurones. The cuticle appears to be secreted by muscle cells, and not distinct hypodermal cells, so these cells are epitheliomuscular cells. This is probably true of other nematodes, e.g. Dorylaimida (Grootaert and Coomans 1980). Though the oesophagus is not generally syncytial, some muscle and gland cells have fused. The number 80 refers to nuclei, not separate cells.

The structure and operation of the oesophagus of two related nematodes, *Rhabditis oxycerca* and *Pelodera lambdiensis*, was studied by Doncaster (1962) using photomicroscopy. The lumen of the posterior bulb expands to form a double, triradiate chamber, termed the haustrulum by Doncaster, because of its pump-like action (see Fig. 2.3). A valve, formed from three flaps, with corrugated cuticular surfaces, bars the entrance of suspended food particles to the haustrulum. Radial muscles in the bulb, when they contract, increase pressure within the bulb, tending to close the lumen ahead of the valve, increasing pressure on its contents. The contraction of these muscles, and the pressure they generate, rotate the flaps towards the posterior, allowing suspended food particles to pass into the haustrulum. The flaps probably grind the food as it passes through. With increasing muscular tension, the lumen of the oesophago/intestinal canal opens, while the passage from the anterior haustrulum to the posterior closes, restricting regurgitation. As muscles relax and intrabulb pressure declines, the pump returns to its resting condition. Further discussion on the operation of the Rhabditid oesophagus will be deferred until Chapter 3. The fine structure of the oesophagus has been described in *P. silusiae* (Yuen 1968a).

In many Tylenchida the mouth stylet leads into a narrow procorpus, followed by a large median bulb, the metacorpus, a narrow isthmus, and a postcorpus, which is largely glandular and which sometimes overlaps the anterior intestine. When plant-feeding tylenchids are feeding, the muscular median bulb alternately pumps the secretions of the dorsal oesophageal gland, which opens by a duct close to the base of the stylet, into plant cells, then pumps liquid food into the intestine. Variations on the structure of the oesophagus with diet and phylogeny have been discussed by Siddiqi (1980), who believes *Hexatylus* may be a primitive tylenchid. In this fungal-feeding genus, the metacorpus is very poorly developed and the ingestion of food depends on the turgor pressure in the fungal hyphae on which it feeds, perhaps aided by rectal pumping. The fine structure of the oesophagus of *Hexatylus viviparus* has been described by Shepherd and Clark (1976b).

The complex musculature of the metacorpus has been described in *Ditylenchus dipsaci* (Yuen 1968b) and *Hoplolaimus* sp (Grootaert, Lippens, Ali, and De Grisse 1976a). In *Hoplolaimus* the metacorpus is formed from 15 cells; six muscle cells, three marginal cells, and six neurones; the isthmus of nine cells. Seymour (1977) constructed a perspex model of the pear-shaped metacorpus of *D. dipsaci*, and used a photoelastic method to analyse stresses in the perspex to simulate stresses in the cuticle during pumping. Radial muscles in the metacorpus, originating from the basement membrane, insert on the triradiate cuticular wall of the lumen. Muscles encircle the middle of the bulb, while the anterior is spongy, and the posterior is largely occupied by the ampullae of the ventral glands. Muscle contraction opens the lumen to a triangular cross-section, setting up stresses within its wall. Stresses concentrate at the apices of the radii, which serve as 'hinges' and Seymour concludes that 'the material of the hinges acts like a rubber spring that stores energy as the radial muscles open the pump lining, and serves to close the lining when the muscles elongate'. The ducts of the two ventrosublateral glands run through the isthmus to open into the oesophageal lumen in the metacorpus, while that of the dorsal gland runs forward to the anterior procorpus.

Stylet protrusion and retraction, which can be rapid and repetitive, has stimulated interest. Three, six, or more protractor muscles, inserted in the head-framework and stylet knobs, are derived from the procorpus, where their nuclei are located. Retraction seems dependent on elastic recoil. In *Tylenchorhynchus dubius* the procorpus consists of a cellular 'core' tapering to the base of the stylet, which encloses the vestibule in a sleeve of tissue. Three marginal cells, containing bundles of microtubules, which may absorb the stresses set up by stylet protrusion, alternate with three protractor muscles (Anderson and Byers 1975). Intracellular fibrils are also present in the procorpus and isthmus of the *Hoplolaimus* sp (Grootaert, Lippens, Ali, and De Grisse 1976*b*). From their dimensions, the fibrils may be actin.

The oesophagus in the tylenchid superfamily Criconematoidea is rather different (see Fig. 2.15). The procorpus and metacorpus are fused to form a short muscular organ, with valved lumen; the isthmus is short, and the glandular postcorpus is reduced. The oesophagus of the many insect-parasitic tylenchids is often poorly developed.

The oesophagus in Aphelenchida resembles that of the plant-feeding Tylenchida superficially (Fig. 2.15), but shows significant differences. Its structure in *Aphelenchoides blastophthorus* has been described in fine structural detail by Shepherd *et al.* (1980). The small stylet leads into a narrow procorpus, from which six protractor muscles are derived, with their nuclei in the metacorpus. The metacorpus is a muscular bulb, with valved lumen, but unlike the Tylenchida the duct of the dorsal oesophageal gland opens in the anterior metacorpus. The isthmus is very short and drawn into the base of the metacorpus, acting as a valve. The nerve ring surrounds the intestine, not the isthmus, encircling the narrow intestinal connection with the isthmus and three large gland cells, which constitute the postcorpus, lie beside the anterior intestine.

In many Dorylaimida and Desmodorida the oesophageal cuticle forms long plates. In *Labronema* sp (Dorylaimida) there are three double plates joined by thinner flexible cuticle permitting the oesophagus to open wide in feeding on its nematode prey, but not to close completely when the oesophagus relaxes, so that glandular secretions can move forward into its prey (Grootaert and Wyss 1978). In *Mononchus aquaticus* (Dorylaimida), which swallows other nematodes whole, the arrangement of its 12 hinged plates allows the cylindrical oesophagus to expand in two stages (see Fig. 2.17) and to close again by muscular contraction (Grootaert and Wyss 1979).

In the Dorylaimida, five glands lie in the posterior muscular portion of the oesophagus, the most anterior dorsal, the four others in pairs in the ventrosublateral sectors (see Fig. 2.12). They can be recognized by their nuclei, with prominent nucleoli, and the openings of the glands into the oesophagus. The relative positions of these glands, important in taxonomy, has been treated systematically by Loof and Coomans (1970). In *Pontonema vulgaris* (Oncholaimidae), a marine scavenger, three large glands are located in the posterior of the cylindrical oesophagus, with ducts that run in the interradial sectors to open at the tips of three teeth (Jennings and Colam 1970).

The oesophagus intrudes into the wider intestine with a cuticle-lined passage for food which also serves as a valve, usually controlled by a sphincter muscle. Inappropriately named the Cardia, its evolution has been discussed by Mapes (1965*a*). In the Dorylaimida it is a complex structure (Clark 1960; Yeates 1972*a*). Seymour and Shepherd (1974) have examined the fine structure of the valve in *Aphelenchoides blastophthorus*

(a)   (b)   (c)

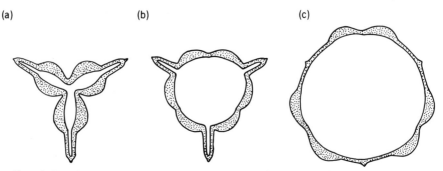

FIG. 2.17.   Schematic representation of the oesophageal lumen cuticular lining in *Mononchus aquaticus*: (a) fully closed; (b) partly dilated, during pauses in the ingestion of prey, with prey retained in the lumen; (c) fully dilated. (Reproduced from Grootaert and Wyss (1979). *Nematologica* **25**, 163–73.)

(Aphelenchida). When the valve is closed, convoluted, closely opposed, interdigitating membranes close the lumen. They suggest that intermolecular forces close the valve until forcibly opened. Similar valves were found in the vas deferens of *A. blastophthorus* and the rectum (see Section 2.9).

2.9. The intestine and rectum

The nematode intestine is usually a simple straight tube, derived from the endoderm, made up of relatively few cells, without intrinsic musculature or specialized glands. In small nematodes the lumen lies between two cells (see Fig. 2.2), while in larger nematodes a number of cells surround the lumen in cross-section. The cells are multifunctional, absorptive, secretory, and store lipid, glycogen, and protein, though there is often functional differentiation along the length of the intestine. The outer surface is separated from the pseudocoelom by a basement membrane. The inner surface possesses microvilli, well shown in electron micrographs of *C. briggsae* (Epstein *et al.* 1971; Vanfleteren 1980). The fine structure of the intestinal cells is similar to that of intestinal cells in many other animals.

The fine structure of intestinal cells of *Tylenchorhynchus dubius* (Tylenchida) has been described by Byers and Anderson (1973) in some detail. The inner surface possesses numerous small microvilli, with central filaments extending into the cell body and small projections radiating from the outer surface of the microvilli. The cells contain vesicles (visible as globules by light microscopy), glycogen, and the typical cell organelles. An unusual feature is bundles of filaments and rod-like elements in paracrystaline arrays (fasciculi), which run through a number of cells. Adjacent cells show lateral junctional complexes at their apical margins of the zonula adhaerens type (terminal bars), but the lateral cell membranes frequently appear incomplete, so that the intestine may function as a syncytium. A tendency to syncytial formation has been observed in a number of nematodes. In the intestine of *Pontonema vulgaris* (Oncholaimidae) (Jennings and Colam 1970), columnar intestinal cells support numerous microvilli, each containing a bundle of filaments. Adjacent cells have interlamellations, but no zonula occludans, zonula adhaerens, or desmosomes were found. Glandular secretion was active in some cells, lipid metabolism in others. Glycogen, lipid, and pigment granules (containing iron) accumulate in the cells. Possibly the cells pass through successive phases of growth, assimilation, and merocrine secretion, culminating in their extrusion into the lumen to disintegrate. Digestion appears to be entirely extracellular.

The rectum, of proctodeal origin, is a short cuticle-lined tube. A rectal sphincter muscle may close the entrance from the intestine, though Seymour and Shepherd (1974) found that in *Thornema wickeni* (Dorylaimida) closure was brought about by aposition at cell membranes without muscles. An anal dilator muscle runs from the dorsal body wall to the anal lip. A number of unicellular rectal glands are commonly present emptying into the rectum. The terminal part of the intestine may also serve as a valve-like region, controlling the passage of fluid into the rectum.

2.10. 'Excretory' system

Nematodes possess organs which have been described as 'excretory' organs, whose function is uncertain, but there is no unequivocal evidence for either an excretory or osmoregulatory function in these organs. After a thorough review of the evidence, Wright and Newall (1976) concluded that 'No quantitative data are available for any nematode excretory site'. Excretion and osmoregulation will be considered in Chapter 3, but it is convenient to label the organs considered in this section as 'excretory' pending a clearer understanding of their role.

In many Adenophorea a large glandular cell, the renette cell, lies ventrally in the anterior pseudocoelom with a duct, which usually opens to the exterior by a ventral cuticular pore, somewhere between the nerve ring and hind end of the oesophagus. The terminal part of the duct usually has a cuticular lining. In *Enoplus brevis* (Enoplida), a marine nematode, its fine structure strongly suggests a secretory function, with secretory vesicles present in the gland and its duct in both sexes (Narang 1970). Some Adenophorea have more than one 'excretory' cell. In *Sabatieria celtica* (Chromadorida), another marine nematode, for example, there are two additional long ventrosublateral gland cells with long ducts which open into the terminal ampulla of the duct of the typical renette cell (Riemann 1977*b*). In *Prionchulus muscorum* and *Longidorus macrosoma* (Dorylaimida), two ventral renette cells open by a single ventral cuticular pore (Jairajpuri and Khan 1975; Aboul-Eid 1969*a* respectively). In *Plectus*, a terrestrial genus, at least three large cells are associated with a long convoluted duct.

In Secernentea the 'excretory' system typically consists of paired lateral canals, associated with the lateral hypodermal chords, and a ventral transverse canal connecting the lateral canals, which gives rise to a short median ventral duct opening to the exterior at the excretory pore, usually near the nerve ring. When the lateral canals extend anteriorly beyond the transverse canal the system is described as an H-system, when they do not, as an inverted U-system. The canals are intracellular, lying within 1–3 cells, with their nuclei associated with the transverse canal. Gland cells may also be associated with the transverse canal, but the distinction between the canal cells and the gland cells is confusing. In *C. elegans*, with an H-system, two gland cells are present (See Fig. 2.18) which contain granules staining

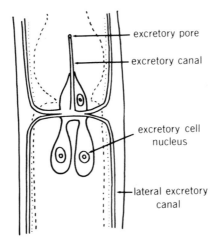

excretory pore

excretory canal

excretory cell
nucleus

lateral excretory
canal

FIG. 2.18.   The cells forming the 'excretory' system in *Caenorhabditis elegans*.

with paraldehydefuchsin, a reaction usually associated with neurosecretion (Mounier 1981).

The fine structure of the system has been described in *Panagrellus*, *Ditylenchus*, and *Globodera* (=*Heterodera*) by Narang (1972). Many microtubules (capitate tubules), in close contact with the hypodermal cell membrane, open into the canals, perhaps suggesting an osmoregulatory function. Similar microtubules have been observed in some animal parasites. The canal system can be stained with a redox dye in *P. redivivus* (Smith 1965). *P. redivivus* has a U-system, while in the two Tylenchids only the left side of an H-system is present (Narang 1972).

A few large cells of unknown function, the coelomocytes, lie within the pseudocoel between the body wall and the alimentary canal. In *C. elegans* four coelomocytes are present in the ventrosublateral pseudocoel.

2.11. Sense organs

Nematodes possess a variety of sensory receptors, described as sensilla (Coomans 1979*b*) (singular sensillum). Free-living nematodes are too small to use electrophysiological techniques to identify the stimuli to which the receptors respond, though microelectrodes have been inserted in *C. elegans* while still alive (Tartar, Stack, and Zuckerman 1977). However, the systematic selection and study of numerous behavioural mutants of *C. elegans*, coupled with studies of the fine structure of their sometimes abnormal sensilla, has opened a new approach to the investigation of the function of sensory receptors (Ward 1973, 1976).

Nematodes are very consistent in the pattern of cephalic sensilla, with

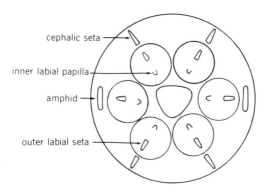

cephalic seta

inner labial papilla

amphid

outer labial seta

FIG. 2.19.   The basic pattern of cephalic sensilla according to De Coninck (1965).

three concentric rings of sensilla on the head surrounding the mouth. This, according to De Coninck (1965), is the primitive arrangement as shown in Fig. 2.19. There are typically six *inner labial* sensilla, six *outer labial* sensilla, four *cephalic* sensilla, and two lateral *amphids*. Quite often the outer labial and cephalic rings combine to form a single ring of 10. The amphids are situated in pockets, while the other sensilla may take the form of papillae, or setae, or be embedded in the cuticle and be invisible externally. The position and form of the cephalic sensilla is important in taxonomy, and their external structure has been illustrated in many taxonomic descriptions, but the study of sections with the electron microscope is required to see their functional morphology. Their fine structure has been reviewed by McLaren (1976*a*, *b*), Coomans (1979*b*), and Wright (1980). Most sensory receptors in nematodes are the dendritic processes of sensory neurones, derived from non-motile cilia, whose perikaryons lie in the central nervous system (Wright 1980). Yet it is only comparatively recently that modified cilia were found in sensory receptors (Roggen, Raski, and Jones 1966), prior to which nematodes were thought to lack cilia or flagella, even in their spermatozoa. Exceptions which show no evidence of derivation from cilia are some ocelli, proprioceptors, and chemoreceptors (Croll, Riding, and Smith 1972; Albertson and Thomson 1976; Lorenzen 1978; Shepherd *et al.* 1980). A variety of different names have been used by various authors for homologous structures and it will be helpful to adopt Coomans's (1979*b*) terminology in future. Coomans lists synonyms in English, German, French, and Dutch for various cells making up the sensillum.

Each head sensillum consists of an extracellular pocket, which may or may not open to the exterior by a cuticular pore, and three kinds of cells (Fig. 2.20). These are a *supporting cell*, a non-neuronal hypodermal cell which secretes a cuticular lining for the neck of the pocket; the *gland* or

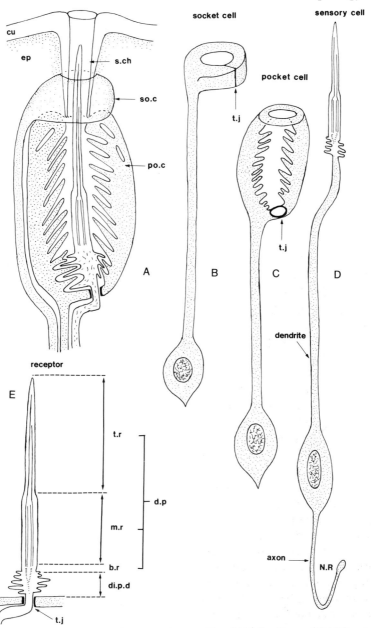

Fig. 2.20. Basic structure of nematode sensilla: (A) distal part with the three components; (B) supporting or socket cell; (C) gland or pocket cell (=sheath cell); (D) sensory cell; (E) ciliary receptor. cu, cuticle; ep, epidermis (=hypodermis); s.ch, sensilla channel; so.c, socket cell; po.c, pocket cell (=sheath cell); t.j, tight junction; t.r, terminal region; mr, median region; br, basal region; di.p.d, distal part of dendrite; d.p, dendritic process. (Reproduced from Coomans (1979*b*). *Rev. Nematol.* **2**, 259–83.)

sheath cell, a secretory cell which surrounds the pocket and secretes into it; and the dendrites of a bipolar *sensory neurone*, which enters the pocket. The dendrites pass through the cytoplasm of the gland cell to enter the pocket. The perikaryons, or nucleated parts, of the supporting cell, the gland cell, and the sensory neurones lie far posterior to the pocket, adjacent to the nerve ring, extending long neurone-like cytoplasmic processes to the pocket. The non-neuronal cells are distinguished from neurones by not forming synapses with other cells. The axons of the sensory neurones, with their perikaryons located more posteriorly than those of the other cells, enter the nerve ring to form synapses.

In *C. elegans* the sensory nervous system has been reconstructed in great detail from photomicrographs of serial sections by computer analysis (Ward, Thomson, White, and Brenner 1975; Ware, Clark, Crossland, and Russell 1975). Ward *et al.* (1975) refer to the supporting cell as a *socket cell*, the gland cell as a *sheath cell*. In *C. elegans* each of the inner labial sensilla has the dendrites of two neurones, each a modified cilium. The pocket leads through a cuticular canal to a minute pore at the tip of a labial papilla, into which is inserted the dendrite of one of the neurones; the other ends within the cuticle. The six outer labial sensilla and the four cephalic sensilla are alike, but not identical. Each has a single neurone, with a single dendrite, which passes through the pocket and ends embedded in the cuticle. Each amphid includes the ciliary dendrites of 12 neurones, several of which branch into two dendrites. Four of the neurones have dendrites of elaborate form, which pass through the amphid sheath cell into the pocket, then re-enter the amphidial sheath cell within its deeply indented surface. The form of these dendrites, though extremely complex, is apparently very similar in different individuals.

Tight junctions between socket, sheath, and neuronal dendrites seal the sensilla pockets from the pseudocoelom. The cell membranes of the sheath, or gland cells, except those of the amphids, have deep lamelliform invaginations from their pockets suggesting secretory activity. The amphidial sheath cells are, however, also clearly secretory. The micro-tubular substructure of the dendrites of the different sensilla depart to a varying degree from the structure of motile kinocilia. In addition to the above sensilla, there are four pairs of neurones with 'non-ciliary' dendrites ending within the head. This description applies to the hermaphrodite; the male has an additional dendrite in each cephalic sensillum, whether a branch from the single neurone or an additional neurone is uncertain, which penetrates the cuticle to open at the surface.

The sensory nervous system of *C. elegans* is bilaterally symmetrical, with a degree of hexaradiate symmetry superimposed. The absence of lateral cephalic sensilla and the presence of lateral amphids is an obvious discrepancy from hexaradiate symmetry. However, another pair of lateral sensilla, the deirids, which in their structure resemble the cephalic sensilla, lie some distance behind the head, with a dendrite penetrating the cuticle, but not reaching the surface. The deirids may be posteriorly displaced cephalic sensilla, which like the cephalic sensilla, but unlike the other head sensilla, possess catecholamines, as evidenced by u.v. fluorescence after formaldehyde fixation (Ward *et al.* 1975).

Neurone-like cytoplasmic processes from supporting cells and sheath cells combine with neuronal dendrites to form six cephalic nerves running to their respective perikaryons located in loose ganglia in the vicinity of the nerve ring. The lateral nerves are larger, because of the more numerous amphidial neurones, which have perikaryons in lateral ganglia. One of the two inner labial neurones synapse with muscle cells, making them sensory motor neurones, whereas the other neurones synapse with interneurones in the nerve ring (Ward *et al.* 1975; Ware *et al.* 1975). The axons from the sensory neurones run directly to the nerve ring, except those of the amphidial nerves which travel by an indirect route to paired ventral ganglia, lying directly posterior to the neuropile of the nerve ring (Ware *et al.* 1975).

The sensory nervous system of other nematodes resembles that of *C. elegans*. The similarity between the arrangement of sensory nerves in *C. elegans* and that in *Ascaris* and *Parascaris*, as worked out by Goldschmidt early this century, is remarkable, bearing in mind the limitations of the light microscope (Johnson and Stretton 1980). The receptors and sensory nerves of *Aphelenchoides fragariae* (Aphelenchida) have been described in comparable detail to *C. elegans* by De Grisse, Natasasmita, and B'Chir (1979). The arrangement is very similar to that in *C. elegans*, though the outer labial sensilla have two dendrites, one ending within the cuticle and one opening by a pore. The amphidial nerves have fewer dendrites. De Grisse (1979), who has described the arrangement of sensilla in a number of other terrestrial Secernentea, found that in some species, the inner, outer, and cephalic sensilla opened by minute pores to the exterior. In Tylenchida, the outer labial dendrites ended within the cuticle, but species varied in whether the cephalic and inner labial sensilla were embedded in cuticle or opened by pores. In Tylenchida, the inner labials may open into the buccal cavity. Other studies of the cephalic sensilla and sensory nerves have been published on Tylenchida by Baldwin and Hirschmann (1973, 1975a, b); De Grisse, Lippens, and Coomans (1974); Endo and Wergin (1977); Rhabditida by Yuen (1968a); Dorylaimida by Roggen, Raski, and Jones (1967), and Lippens *et al.* (1975).

Nematodes differ in the numbers of dendrites associated with some sensilla, tending to be more numerous in Adenophorea than Secernentea (Coomans 1979b). The amphids of Adenophorea are more complex than those of Secernentea, involving two chambers, the inner more posterior

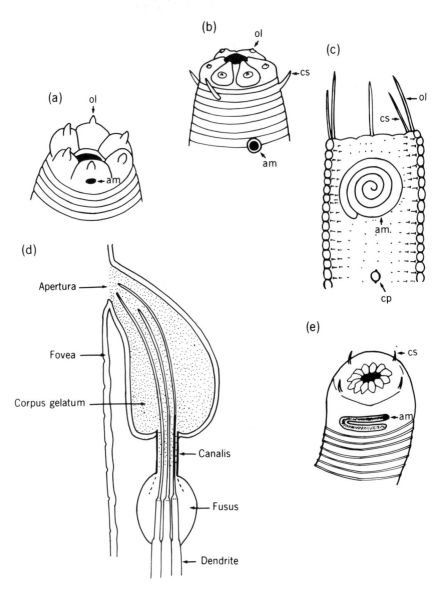

FIG. 2.21. Examples of the location and external opening of the amphids: (a) opening on the lateral lips of *Panagrolaimus*, as in many Secernentea; (b) to (c) opening posterior to the lips, as in many Adenophorea; (b) *Plectus*; (c) *Pomponema*; (d) *Actinonema*; (e) diagrammatic section through amphid of *Tobrilus* (From Storch and Riemann (1973).) am, amphid; ol, outer labial papilla; cs, cephalic seta; cp, cuticular pore.

being the amphidial pocket, see Figs. 2.20, 2.21, and Plates 1 and 2. The structure has been described in representatives from eight families by Riemann (1972) as seen by Normarski interference light microscopy. The amphid pocket, termed by Riemann the *fusus*, contains the ciliary dendrites and communicates by a narrow canal, *canalis*, with a cuticle-lined invagination, the *fovea*. Terminal filaments of the sensory dendrites, which stain intensely with Alican blue, extend through the fovea to the aperture, *apertura*. It is the fovea and apertura which may take diverse forms in various Adenophorea. The fovea is filled with a gelatinous secretion, *corpus gelatum*. The fine structure of an Enoplid, *Tobrilus aberrans*, displaying all these features, was described by Storch and Riemann (1973). The 16 dendrites, with $9 \times 2 + 10$ microtubule formula, lose this substructure in the canalis.

Comparison between nematode receptors and those better understood in other invertebrates suggest that the sensilla with dendrites ending within the cuticle are probably mechanoreceptors, e.g. the external cephalic sensilla. Those with an opening to the exterior may be chemoreceptors, and studies of behavioural mutants support a chemosensory role for the amphids (Ward 1973). This is further reinforced by the fine structure of the dendrites. The inner labial sensilla in *C. elegans* may be both mechanoreceptors and chemosensory. However, it seems that homologous sensilla may serve different functions in different nematodes (Coomans 1979b). The extra male dendrites in the cephalic sensilla are probably chemosensory on behavioural grounds. *Aphelenchoides blastophthorus* possesses two different receptors asscociated with the stylet (Shepherd *et al.* 1980). A 'gustatory' neurone encircles the stylet shaft where the shaft is penetrated by fine 10-nm canals. A neurone ending packed with 500 parallel microtubules lies adjacent to the stylet; presumably a mechanoreceptor.

Free-living nematodes may respond to light without apparent photoreceptors, but paired eye spots are common in aquatic nematodes (see Fig. 2.24). These are located in the anterior oesophagus and are associated with photoreceptors, which may be of two kinds, rhabdomic or ciliary. The pigments, located within vesicles in oesophageal cells are generally melanins (Croll, Evans, and Smith 1975; Viglierchio and Siddiqi 1974), though in the larvae of the insect parasite *Mermis subnigrescens* the pigment of the chromotrope appears to be a haemoglobin (Ellenby 1964; Croll *et al.* 1975). In *Enoplus communis* eye spots are present, but a photoreceptor could not be found by Croll *et al.* (1975), so the presence of 'eye spots' may not invariably be associated with photoreception. In other nematodes the presence of an associated lamelliform rhabdome is strongly suggestive of photoreception. Such a receptor has been described in *Chromadorina bioculata*, a freshwater chromadorid, by Croll *et al.* (1972).

A rhabdome lies adjacent to each of the paired pigment spots in such a way as to shield the rhabdome from light from one side of the animal. The neuronal connections were not elucidated. Earlier, Croll (1966*a*, *b*) had described the spectral photosensitivity. In *Deontostoma californicum* (Enoplida) each ocellus consists of a pigment cell, which partially encloses and screens from behind concentric lamellae (the rhabdome) within a sensory receptor (Siddiqi and Viglierchio 1970). The perikaryon of this bipolar sensory neurone forms the lens; the axon enters the adjacent lateral cephalic nerve.

A different kind of ocellus is found in *Oncholaimus vesicarius* (Enoplida), with a ciliary receptor instead of a rhabdome (Burr and Webster 1971). Pigment vesicles in the anterior oesophageal cells partially shield the amphids on either side. The amphids appear to be bifunctional (Burr and Burr 1975); three of the sensory neurones, with 28–36 dendrites, appear to be chemosensory, with their dendrites extending out of the pocket. The dendrites of the fourth neurone, about 10, project medially around the front of the pigmented region of the oesophagus. The dendrites differ in their substructure, those of the former three neurones resembling olfactory dendrites in other animals. Burr and Burr (1975) observed negative phototaxis in this nematode, and suggest the repeated turning of the head as it moves would allow alternate exposure and shielding of the photoreceptor from light.

The presence of two entirely different kinds of photoreceptors within the Nematoda is of interest because of previous suggestions that these two types of photoreceptors may characterize different phylogenies. The evidence that either sensillum is a photoreceptor is based on behavioural and structural evidence. Electrophysiological records of the response of the receptor cell to light are lacking.

Sensory receptors occur along the body of nematodes and are associated with the reproductive organs (see Section 2.13). In *Aporcelaimellus* (Dorylaimida) a number of sensilla, innervated by dendrites from the lateral cephalic nerves, open by lateral pores behind the amphids (Lippens *et al.* 1975). In *Chromadorina germanica* (Chromadorida) rows of body pores are each associated with a merocrine hypodermal gland cell and a bipolar sensory neurone (Lippens 1974). The secretions from the gland and two ciliary dendrites from the neurone enter the duct of the gland. An elongated glial cell lies on the short axon which leads to the lateral chords. Epidermal glands are also associated with bipolar neurones in *Thoracostoma* (= ? *Deontostoma*) *californicum* (Enoplida). Setae along the body of the nematode contain dendrites of sensory neurones (Maggenti 1964).

The Enoplida possess sensilla, not present in other nematodes, which from their structure are almost certainly stretch receptors (Lorenzen 1978, 1981*a*). They lie in series beneath the cuticle of the lateral fields, either orientated along the anterior posterior axis, or obliquely across the lateral fields, called metanemes by Lorenzen (1981a) (see Fig. 2.22). Six mechanosensory neurones have been located in *C. elegans* by laser beam surgery with anterior processes running adjacent to the cuticle in the hypodermis. They can be recognized electronmicroscopically by their characteristic microtubules and extracellular 'mantels' (Chalfie and Sulston 1981). Touch insensitive mutants with defective neurones have been isolated.

2.12. The nervous system

The central nervous system of nematodes takes the form of a circum-oesophageal nerve ring and associated ganglia. Six sensory nerves, subdivided distally into 22 or more branches, run from these ganglia along the hypodermal tissue to the head sensilla. Dorsal and ventral motor nerves run posteriorly from the nerve ring, adjacent to the dorsal and ventral hypodermal chords. Posteriorly there is a pre-anal ganglion associated with the ventral chord, and in males additional ganglia associated with the copulatory apparatus. In nematodes extensions from the muscle cells reach the nerve ring and the motor nerves, where they form synapses with motor neurones, instead of the muscles receiving axons from the motor nerves, as in other animals. An intrinsic oesophageal nervous system is connected to the nerve ring by interneurones. The most complete description of the nervous system comes from the study of serial electronmicrographs of *C. elegans*. The oesophageal nervous system has

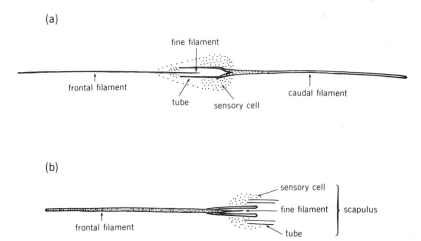

FIG. 2.22. Metanemes, stretch receptors arranged in series beneath the lateral cuticle of many Enoplida: (a) example with caudal filament; (b) example without caudal filament. (Reproduced from Lorenzen (1978). *Zoologica Scripta* **7**, 175–8.)

already been described in Section 2.8, and the sensory neurones associated with the head sensilla have been discussed in Section 2.11. A schematic reconstruction is shown in Fig. 3.5.

The nerve ring receives sensory inputs from the head sensilla, the oesophagus (by two interneurones), and the head musculature. Its output is primarily to the musculature of the body wall. The sensilla show hexaradiate symmetry, the musculature fourfold symmetry, and the central nervous system a high degree of bilateral symmetry. Chemical synapses and gap junctions are recognizable in the nervous system.

The nerve ring and its neuronal connections with sensory receptors and head musculature has been described by Ware *et al.* (1975). The nerve ring consists of a neuropile of interneurones and nerve processes, which tend to run circumferentially, with numerous synapses by which it receives sensory input from associated ganglia. Four kinds of synaptic vesicles are recognizable. Point synapses are the most common type, with gap junctions, which are thought to indicate electrical coupling between cells, frequent, especially between muscle processes. The nucleated cell bodies, the perikaryons, of the sensory neurones are clustered in ganglia at the front of the nerve ring, one ganglion associated with each of the six anterior sensory nerves in an arrangement corresponding to the arrangement of sensilla on the head. The amphidial perikaryons, however, lie in lateral ganglia, their axons entering the nerve ring via the ventral ganglia. Lateral and a posterior ventral ganglia can be recognized in the loose cluster of ganglia adjacent to the nerve ring.

The anterior 32 muscle cells send slender processes into the nerve ring which spread out as thin sheets within the neuropile of the ring. The sheets correspond strictly in position to the location of the muscle cells, four quadrants of eight, four posterior and four anterior in each quadrant. The muscle processes form frequent gap junctions between one another and chemical synapses with interneurones. The four posterior muscle cells of each quadrant also send processes to the corresponding dorsal and ventral cords.

The ventral and dorsal nerve chords in *C. elegans* have been described in great detail by White *et al.* (1976). The ventral chord contains 57 motor neurones which innervate both the dorsal and ventral body-wall muscles. Motor neurones which innervate the dorsal muscles do so by commisures running circumferentially to the dorsal chord. The ventral chord originates within a retrovesicular ganglion of about 20 cells, which is separated from the ventral ganglion of the nerve ring by the excretory pore, and ends in the pre-anal ganglion.

There are five classes of motor neurones, distinguishable morphologically and by their synaptic connections. Within each class, the neurones are arranged in sequence along the ventral chord, with sharply separated fields of synaptic connections, so that at any point along the cord only one neurone of each class forms synapses. Neurones of the same class are exclusively interconnected by gap juctions with neurones of the same class. The dorsal chord receives the axons of motor neurones in the ventral chord by lateral commisures, which then run anteriorly or posteriorly in the chord, depending on the class to which they belong. The ventral chord also has four classes of interneurones, distinguishable by their synaptic connections, which may be either chemical synapses or gap junctions.

Muscle cell processes, interconnected by gap junctions, come into aposition with the ventral or the dorsal chord, according to their dorsal or ventral location, where myoneural synapses occur. Four muscles in each quadrant are innervated from both the nerve ring and either the dorsal or ventral chords. The motor neurone classes differ in their polarity, i.e. the direction from dendrite to axon with reference to the anterior posterior axis of the body, and the polarity of the same class is reversed in the two chords. Presumably this is related in some way to the difference in the phase of the waves of contraction propagated along the dorsal and ventral muscles during locomotion. The nervous system of *Ascaris suum* shows intriguing similarities to that of *C. elegans* (Johnson and Stretton 1980), but the significance of the repeating patterns of different classes of neurones along the dorsal and ventral chords remains to be worked out.

2.13. Reproductive and accessory sexual organs

The reproductive organs of both sexes are basically similar, consisting of epithelial tubes lying within the pseudocoelom, in which gametogenesis is usually restricted to the closed tips of the organ. Most nematodes are bisexual, though parthenogenesis, and a form of self-fertilizing hermaphroditism, in which first spermatozoa then ova are produced in the same gonad is common (Maupas 1900). Abnormal intersexes are not infrequent. Fertilization is internal following copulation. The full development of the gonads and accessory organs occurs late in larval life from two pairs of primordial cells, a germ cell and duct cell in each pair. A discussion of gametogenesis will be deferred until Chapter 6.

In the male there is usually a single gonad, i.e. a monorchic condition, though diorchicism is not uncommon, as in *Mononchus* (see Fig. 4.1) where two testes open into a single terminal gonoduct. The germinal zone, or testis, is followed by a storage region for spermatozoa, the vesicular seminalis, with a secretory epithelial lining, sometimes with a short intervening region, the vas deferens. Muscle cells, associated with the vesicular seminalis, in some form an ejaculatory organ, and there may also be unicellular glands associated with the gonopore. The male gonoduct almost always opens close to the anus into a cloaca. There are usually two spicules, hollow cuticular rods, often of complex form, lying in a spicule pouch, dorsal and posterior to the gonopore. Retractor and protractor muscles, inserted on the head of the spicule, allow these to be protruded from the cloaca and inserted into the female vulva during copulation, aiding in the transfer of spermatozoa. A cuticular thickening in the dorsal wall of the spicule pouch, the gubernaculum, is usually present, sometimes also of complex shape with associated muscles.

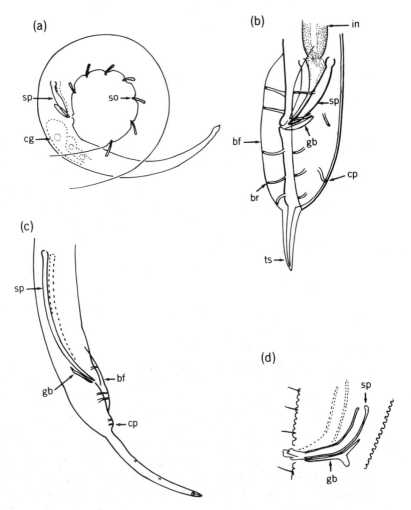

FIG. 2.23.   Male copulatory organs: (a) *Aphanolaimus*; (b) *Rhabditis oxycerca*; (c) *Diplolaimelloides* (d) *Gonionchus*. sp. spicule; so, supplementary organ; in, intestine; gb, gubernaculum; bf, cuticular flap of copulatory bursa; br, bursal 'ray' or bursal sensory papilla; cp, sensory papilla external to bursa; ts, tail spike.

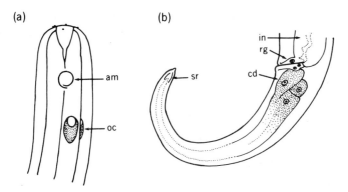

FIG. 2.24. Ocelli and caudal glands: (a) *Diplolaimelloides*; (b) *Plectus*. am, amphid; oc, ocellus; in, intestine; rg, rectal glands; cd, caudal glands; sr, spinnerette.

The ciliary dendrites of sensory neurones have been shown to run to the tip of the spicules, where there is a minute pore opening to the exterior in several species. Sensory papillae are also located on the cuticle in the vicinity of the cloaca in most nematodes. In many Adenophorea, a ventral row of sensory papillae extends forward from the cloaca, the supplements or supplementary organs (see Plate 3). Some nematodes have cuticular folds on both sides of the cloaca, the bursa copulatrix, which encloses the female vulva during copulation. Examples of the male copulatory organs are illustrated in Figs. 2.23 and 2.24. The fine structure of the spicules and associated structures has been described in several Tylenchida (Clarke, Shepherd, and Kempton 1973; Wen and Chen 1976; Natasasmita and De Grisse 1978). Copulation and insemination has been described in *Panagrellus redivivus*, Rhabditida (Duggal 1978*b*) and *Ditylenchus myceliophagus*, Tylenchida (Cayrol 1970).

The female nematode usually has two gonoducts, i.e. didelphic, which usually open by a mid-ventral vulva. Monodelphic species are also common, when the vulva tends to be more posterior. The two gonoducts often extend in opposite directions from the median vulva, twisting around the intestine, with their terminal germinal zones turned back upon themselves, i.e. reflexed; whilst in monodelphic species, the gonad may take an elongated S form, turning back on itself more than once, with the second vestigial gonoduct represented by a post-vulval sack. The female gonoduct lacks extrinsic musculature and the movement of eggs is brought about by general body movement, though there may be one or two sphincter regions, formed from groups of cells which close the duct so as to retard the movement of ova. Extrinsic muscles are often associated with the terminal uterus and vulva, aiding in oviposition. Examples of specific descriptions of the female gonoduct in Tylenchida can be found in Geraert (1972, 1976, 1980); of Dorylaimida in Coomans (1964) and Geraert, Grootaert, and Decraemer (1980); and of Rhabditida in Geraert, Sudhaus, and Grootaert (1980). Its fine structure is described in *Aphelenchoides myceliophagus* (Aphelenchida) by Yuen (1971).

The cellular structure of the female gonoduct has been reviewed by Geraert (1980), on whose general conclusions I rely here. Its structure is much better known in Tylenchida than in other nematodes, but sufficient work has been done on other orders to draw some general conclusions, though there are difficulties in recognizing which regions are homologous in different orders. Within a species its cellular constitution, involving 30–100 cells, is highly constant. The ovary consists of a germinal zone, where oogenesis occurs, followed by a growth zone in which oocytes increase in size, and a maturation zone in which yolk may accumulate in the ova. The ovary is followed by a narrow region, the oviduct, whose cellular constitution seems to characterize different orders. In Enoplida and Dorylaimida it is formed from a single column of cells, numbering two or three to 36, without enclosing any evident duct lumen. In Tylenchida and Rhabditida there are two columns of cells ranging from a single pair to about 50. In some Adenophorea there are three rows of cells. The oviduct leads into an expanded tube surrounded by cells, the uterus, in which fertilization occurs and the egg shell is laid down. Frequently, the first portion forms a distinct spermatheca, in others a specialized fertilization zone. In some Oncholaimidae (Enoplida) a supplementary duct, the Demanian system, opens into the closely associated intestinal lumen to serve as a spermatheca (Rachor 1969). The structure of the female gonad in *C. elegans* is illustrated in Fig. 2.25.

Nematode eggs when fully developed possess a multi-layered shell, but may none the less become highly constricted passing through the vulva (Brun and Cayrol 1970). The shell has been most fully described in *Ascaris lumbricoides* (Foor 1967), and four species of Tylenchida (Bird 1968; Bird and McClure 1976). There are three layers, a very thin outer vitelline layer, which in *Ascaris* separates immediately after fertilization, a chitinous layer, and an inner lipoidal layer, all formed from the egg. In *Meloidogyne javanica*, in which the eggs develop parthenogenetically, the vitelline layer develops from a thickening of the plasmalemma as the eggs pass through the spermatheca. The chitin and lipid layers also contain much protein, probably as complexes, the chitin layer conferring mechanical strength, the lipid layer impermeability to many polar compounds. Amino acid analyses show the eggs to be rich in proline.

Nematode spermatozoa are unlike those of most other animals in being non-flagellate and lacking an acrosome. Those of parasitic nematodes, especially Ascarids, have received the most attention (see review by Foor

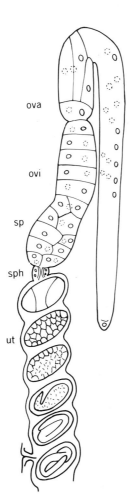

ova

ovi

sp

sph

ut

FIG. 2.25. The female gonad in *Caenorhabditis elegans*. ova, ovary; ovi, oviduct; spt, spermatheca; sph, sphincter; ut, uterus. (Reproduced from Geraert, Sudhaus, and Grootaert (1979). *Annls Soc. r. zool. Belg.* **109**, 91–108).

1970). In the mature spermatozoa, microtubules are not evident and the nucleus lacks a membrane. The spermatozoa are motile, though not necessarily before being deposited in the female gonoduct, possessing an anterior pseudopodium, which in the absence of an acrosome, may also facilitate fusion with the ovum. In some nematodes, spermatozoa may move forward in short chains, as in *Panagrellus redivivus* (Duggal 1978c) and *Ditylenchus destructor* (Anderson and Darling 1962). In *Acanthonchus calathrus* (Chromadorida) they may possess sharp spikes, possibly serving to anchor them on the female gonoduct (Gerlach, Schrage, and Riemann

1979), though in most nematodes they appear to adhere to the sperma-
thecal epithelium. Fine structure has been described in *P. silusae*
(Pasternak and Samoiloff 1972) and *Rhabditis pellio* (Beams and Sekhon
1972). Final maturation may not occur until in contact with the female
gonoduct.

2.14. Morphometrics

Measurements are important in taxonomy and this aspect will be discussed
in Chapter 9, but may also, when critically analysed, elucidate relationships between form and function. Geraert (1978*a*, *b*, 1979*a*, *b*) has
analysed the relative rates of growth of body length and breadth,
oesophageal length, position of the oesophageal bulbs, stylet length, and
vulval position throughout post-embryonic growth in many nematodes,
taking the data from published measurements. Graphical representation of
the relative rates of growth of two dimensions gives a clear visual
representation of changes in proportions as the nematode grows. In many
cases the relationship can be expressed as a quadratic equation. Yeates
(1973*c*) used logarithmic equations to describe the relationship between
biomass and development stage in nematodes from several taxa, as well as
to derive allometric ratios for the relative rates of growth of several
structures.

An analysis of 42 species of Tylenchida (Geraert 1978*a*) shows that the
growth in length of the oesophagus does not keep pace with the growth in
the length of the body. However, within a genus, the position of the
median bulb along the oesophagus remains fairly constant. The further
posterior a median bulb is placed, the more powerful it tends to be. Longer
stylets are associated with a shorter pre-bulb oesophagus, shorter stylets
with a longer pre-bulb oesophagus. Increasing glandular tissue in the
post-bulbar region leads to an overlap of the intestine by glandular tissue,
which does not lengthen the post-bulbar oesophageal lumen.

In Dorylaimida (Geraert 1979*b*) the body generally becomes relatively
thinner as the nematode grows. In those taking liquid food from punctured
plant cells the anterior oesophagus is not muscular, and the growth of the
oesophagus is restricted, while in those taking particulate food, such as
predators, the whole cylindrical oesophagus is muscular and its growth in
length continues throughout development. Yeates (1979*a*) has also
analysed the relationship between oesophageal length and body length in
the Dorylaimida and several Tylenchida.

When parasitic nematodes from the Ascaridida, Strongylida, Spirurida,
and Mermithidae are taken into account as well as free-living nematodes
(Geraert 1979*a*), it becomes apparent that an oesophagus with one or two
muscular bulbs can be shorter than one without, but requires an increase in

body width. Highly motile nematodes remain long and thin, while inactive nematodes can be relatively much broader. Changes in the form of the oesophagus during the development of many parasites can be explained by the changing demands for motility and reproduction, the latter requiring a great increase in size in the female. The relationship between tail lengths and body lengths, and the position of the vulva along the body has similarly been discussed by Geraert (1979b).

Roggen (1970a) approached the relationships between oesophageal dimensions and the size of the body from an 'engineering' point of view. Having made some rather far-reaching assumptions and extrapolations, Roggen calculates that there is a minimum length for the oesophagus to function effectively, unless the internal pressure of the body is lowered, and that this will interfere with the characteristic nematode method of locomotion. It is assumed that closure of the oesophagus depends on the hydrostatic pressure in the pseudocoel. Interestingly, the smallest nematodes, less than 0.3 mm in length, belong to families with atypical locomotion, lending credence to the hypothesis that typical nematode morphology is inefficient in animals less than 0.3 mm long. Roggen suggests a simplified model in which the muscular component may be independent of body length, while the glandular part may increase with the 2/3 power of the body length. Yeates (1979a) found evidence for such a relationship in 17 genera of Dorylaimida and Tylenchida, but not in *Dorylaimellus*. Roggen (1970b) argues that the upper limit to the diameter of the oesophagus is set by diffusion. A theoretical treatment of the mechanics of the nematode oesophagus by the same author (Roggen 1973, 1979) has already been referred to (see Section 2.8).

3. Physiology

3.1. Introduction

NEMATODES, when compared with other large phyla, show remarkable structural uniformity, which is associated with similarities in their methods of locomotion and mechanical and skeletal systems. There are exceptions to these generalizations, but the success of nematodes in inhabiting such a wide range of environments rests on their physiological, behavioural, and biochemical adaptations, not on their structural diversity. The mechanics of locomotion was investigated by Gray, Wallace, and others over 20 years ago, but has received relatively little attention since then. In contrast, great progress has been made in studies of nematode neurobiology and behaviour. I have used the term behaviour in preference to ethology because the work I shall describe, involving attraction, repulsion, aggregation, copulation, and feeding has been interpreted in neuromuscular and sensory physiology terms rather than in the ethological concepts more usually applied to animals with more complex brains.

Terrestrial and freshwater nematodes show remarkable powers of osmoregulation, but lack recognizable osmoregulatory organs. Many free-living nematodes live in environments where oxygen may be deficient, but very little is known about their capacity to respire anaerobically, despite the evidence for well-developed anaerobic respiration in parasitic nematodes. I have considered the factors determining the availability of oxygen to nematode tissues, and the responses of nematodes to lack of oxygen, in this chapter. Discussion of respiratory metabolism will be deferred until Chapter 5. Growth and respiration are temperature-dependent and studies have defined these relationships as well as demonstrating interesting biochemical and behavioural adaptations to differing environmental temperatures.

The capacity of many nematodes to tolerate desiccation, freezing, high temperatures, oxygen deficiency, and oxygen stress, in a quiescent or cryptobiotic condition, enables many to occupy habitats where such conditions occur, opening up large areas of the earth's surface to terrestrial nematodes, though all nematodes require a film of moisture for active life. Studies of cryptobiosis have given valuable insights into the physiological adaptations involved.

3.2. Locomotion

Nematodes typically propel themselves by undulatory movements brought about by waves of contraction passing along the body wall musculature.

Undulations come from the action of muscles on the internal pressure of the body fluid and the elastic properties of the cuticle. With minor morphological and physiological variations, undulatory propulsion enables different nematodes to move through soil, sand, mud, and the tissues of plants and other animals. Undulatory movements also allow nematodes to swim, though not very efficiently. This typical nematode method of locomotion is associated with characteristic muscle and cuticle morphology and is fundamental to the nematode body plan.

A few nematodes show quite different methods of locomotion, with corresponding modification of the form of the body and cuticular structure (Stauffer 1925). Undulatory propulsion involves the propagation of waves of contraction along the dorsal and ventral musculature, 180° out of phase, so that as dorsal muscles contract, the opposite ventral muscles relax. The Criconematidae are untypical in that the waves of contraction in the opposite dorsal and ventral muscles are in phase, so that the nematodes progress through the soil like earthworms, aided by strongly developed annulations in the cuticle, often with backwardly directed spines. Marine Epsilonematidae and Draconematidae loop along, reminiscent of some caterpillars, the latter using glandular setae to cling to surfaces. Desmoscolecidae, mostly marine, move on the tips of articulated setae as contractile waves move along the body.

These methods of locomotion are the exceptions, and most nematodes have a similar limited repertoire of body movements (Crofton 1971):

1. Head oscillations without forward progression;
2. Slow backward body waves used in crawling forward;
3. Fast backward body waves used in swimming;
4. Slow forward body waves used in crawling backwards;
5. Stationary long-period body waves, bracing the body against the substratum;
6. Shortening the body;
7. Coiling the body into a flattened spiral or helix (when inactive);
8. Coiling of the male around the female in copulation.

Calcoen and Roggen (1973) suggest that the body undulations do not follow sine curves, but are probably an elastica, the shape of a cylinder bending under axial compression, because the elastica minimizes the work done in bending. They found close agreement between an elastica and the body waves of *Enoplus communis*; *Rhabditis* sp., and *Turbatrix aceti* recorded on videotape. The angle θ between the tangent at a point on the body and the axis of movement was a sine function of the distance along the body. An equation of the form $\theta = k \sin \alpha L$, where L is measured along the body, instead of along the axis of movement, may be logical, because of the way muscular contraction is propagated.

(a)

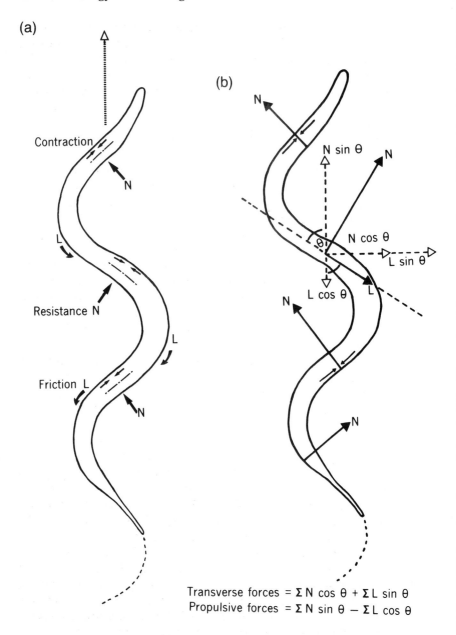

Contraction

Resistance N

Friction L

(b)

N sin θ

N cos θ

L sin θ

L cos θ

Transverse forces = $\Sigma N \cos \theta + \Sigma L \sin \theta$

Propulsive forces = $\Sigma N \sin \theta - \Sigma L \cos \theta$

FIG. 3.1. The forces acting on a nematode crawling through a homogeneous medium: (a) regions where body-wall muscles are contracting and reactions of substratum to movement of nematode; (b) resolution of reactions to forces exerted by nematode on medium into transverse forces, which cancel out, and propulsive force. (Adapted from Gray (1968).)

Many animals propel themselves by undulatory propulsion. In vertebrates such as eels and snakes, shortening of the body, when muscles contract, is resisted by the spinal column. In nematodes, bending of the body through the asymmetric contraction of the body wall musculature depends on internal fluid pressure to develop a bending couple. Muscular contraction will tend to shorten the body, increasing the hydrostatic pressure, which can be considered to act at the geometrical centre of the cross-section, or centroid. Thus a couple, Fb, develops from the localized tensile force of contraction F acting at a distance b from the centroid (Alexander 1979). Shortening of the body would increase the cross-sectional area, stretching the cuticle. Thus the complex fibres and plates found in nematode cuticles serve to resist stretching, maintaining internal pressure and facilitating bending.

In moving through soil, body waves exert forces on the soil particles, the resultants of which propel the nematode forward. Such movement is most readily analysed in a homogeneous transparent agar gel. In moving through stiff agar, or on its surface where they incise shallow grooves, the track of the nematode follows smoothly the path of the body waves. The mechanics of undulatory propulsion in nematodes has been analysed by Gray and Lissmann (1964) and more fully discussed in Gray's (1968) monograph on animal locomotion.

The forces acting on a nematode moving through agar are illustrated diagrammatically in Fig. 3.1. In steady motion, the lateral forces acting on the body tend to cancel out, while the sum of the components propelling the nematode forward equals the sum of frictional forces resisting forward movement. In the diagram, muscle contraction is shown as exerting forces on the substratum where the body wave cuts the line of advance. Increasing angle θ would increase the forward component of the opposing force, so that to increase the amplitude and decrease the wavelength of body waves would compensate for any increased friction. Observations show that living nematodes respond to an increase in resistance in this way.

Nematodes can move in thin films of water, a little shallower than the body diameter, because their body waves distort the film so that the resultant of asymmetric surface tension resists bending of the body (Wallace 1959c) (see Fig. 3.2). When the film is much thinner than the body, movement is restricted; when deeper, crawling becomes swimming (Wallace 1958a). Wallace (1958a, b) has studied movement by soil nematodes in thin films (*Heterodera schachtii* and *Ditylenchus dipsaci*). When the speed of movement over the substratum increases, wavelength increases and amplitude decreases. *Hoplolaimus indicus* shows similar relationships (Azmi and Jairajpuri 1976). Mathematical analysis would probably show that, as the speed increases, decreasing amplitude and increasing wavelength optimize propulsive power. However, the particular

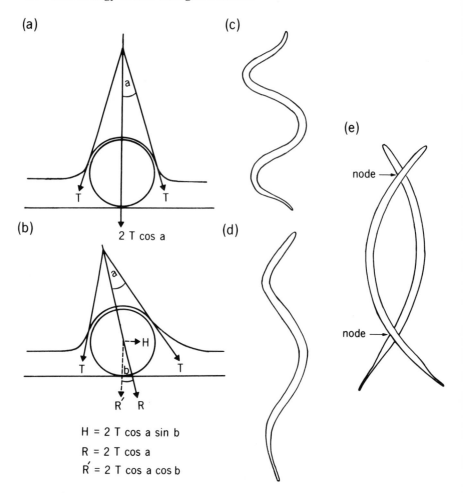

FIG. 3.2.   (a) and (b) The action of surface tension on a nematode crawling in a thin film of water showing how surface tension opposes lateral movement of the body (adapted from Wallace (1959c)); (c) and (d) Changes in the form of the body waves with increasing depth of water and decreasing resistance; (d) with deeper water; (e) A nematode swimming showing nodes, where lateral displacement of body is minimal.

relationship between wavelength, body length, and maximum speed differs in different species (Wallace 1958b). Generally, soil nematodes move most rapidly when their length is about three times that of the diameter of soil particles.

When nematodes are immersed in water deeper than their diameter, they can swim; propulsion depending on the resistance offered by the

TABLE 3.1

Crawling on agar and swimming in water by two free-living nematodes
(Gray and Lissmann 1964)

Species	Panagrellus silusiae		Turbatrix aceti		Units
	Crawling	Swimming	Crawling	Swimming	
Length ℓ	962		840		μm
Max. dia.	19		28		μm
Wave frequency f	0.85	3.0	0.63	5.2	waves s^{-1}
Wave length w	452	770	277	714	μm
Wave speed V_w	512	2200	312	3700	μm s^{-1}
Ratio: amplitude/w	0.21	0.23	0.22	0.15	
Ratio: w/ℓ	0.47	0.80	0.33	0.85	
Distance moved/wave	485	213	429	138	μm
Speed of nematode V	412	640	270	718	μm s^{-1}
Slip $(1 - V/V_w)$ 100	20	71	14	81	per cent

Measurements based on about 20 specimens.

water to body waves, the water being propelled in the opposite direction. The relative importance of inertia and viscosity when a body moves through water is expressed by the Reynolds number. With nematodes, because of their small size, this number is small, so that viscosity is of much more importance than inertia (Gray 1968). When swimming, the rate of propagation of body waves V_W exceeds the rate of movement over the ground V_X and there is slip

$$\text{Slip percentage} = [(V_W - V_X/V_W)] \times 100.$$

Examples of these values in *Rhabditis* sp are given in Table 3.1, and a comparison between crawling and swimming in Table 3.2. Wallace and Doncaster (1964) analysed the swimming of 14 species of nematodes, parasitic and free-living, using cinematography. Most species swim faster than they crawl, and there is usually only a single asymmetric wave present at a time. Cinematography showed two nodes, one near each end of the

TABLE 3.2

Rhabditis sp gliding between starch grains

Wave frequency per s f	Wave length (μm) λ	Speed of waves V_w (μm/s) $V_w = f\lambda$	Speed of progression (μm/s) V_x	$\dfrac{V_x}{V_w}$	Slip (per cent)
0.66	650	429	260	0.60	40
1.04	600	624	520	0.85	15
1.80	700	560	420	0.75	25
0.55	850	467	330	0.71	29
0.69	750	517	220	0.42	58
0.57	470	270	240	0.89	11

Reproduced from Gray and Lissmann (1964).

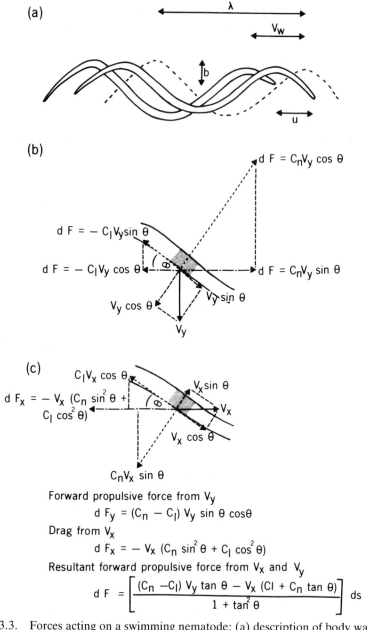

F$_{IG}$. 3.3.   Forces acting on a swimming nematode: (a) description of body waves; (b) resolution of the forces reacting to the component of the movement of a small element of the body at right angles to the axis of forward movement; (c) resolution of the forces reacting to the component of movement along the axis of forward movement. (Adapted from Gray (1968).)

body at which lateral displacement was minimal (Fig. 3.2). (These can often be seen by eye with the microscope in fast-swimming species.) They found that they could express forward speed as a function of body length and wave frequency

$$\text{Speed (mm min}^{-1}) = 0.147L \text{ (mm)} \times f \text{ (waves min}^{-1})$$
$$\text{at } 18°C \ (r = 0.96).$$

To swim by undulatory propulsion the resistance of the water to lateral movements of the body C_N must be greater than that to tangential movements C_L (Fig. 3.3). Gray and Lissmann (1964) attempted to verify experimentally the theoretical hydrodynamic analysis of long cylindrical animals moving through water. Treating a nematode as a long cylinder falling through fluid gives a theoretical value of $C_N/C_L = 2$, which suggests 80 per cent slip. However, nematodes generally swim with less slip. The authors observed that swimming nematodes generate turbulance, unlike the smooth flow over a sinking cylinder.

The forces generated by flagellae have been more thoroughly analysed mathematically. Those which generate sinusoidal waves in one plane, are comparable to swimming nematodes (though with smaller Reynolds numbers). Theoretically, for a given speed, there is an optimal relationship between wave amplitude and wavelength which will minimize the power required. Such a relationship, as already noted has been observed in nematodes moving in thin films of water. Animals swimming by undulatory propulsion must control tendencies to yaw from side to side, either by fins or asymmetry in the body waves which usually increase in amplitude as they pass down the body. Body waves shorter than body length will tend to cancel out the effect of transverse forces acting on the body. Gray (1968) concluded that both these factors were important in *Turbatrix aceti* which swims actively in its natural habitat. However, many nematodes swim with body waves less than body length, and the amplitude in some decreases as they pass along the body. The lateral fields of some aquatic nematodes form keels (Riemann 1976) which may stabilize the nematode during swimming, perhaps preventing rolling, and the post-anal tail perhaps plays some part in stabilization.

3.3. Feeding and ingestion

The oesophagus is a muscular pump, equipped with valves, which sucks food in through the mouth and forces it into the nonmuscular intestine. In some nematodes the secretion of the oesophageal glands is pumped out of the mouth into the food between periods of ingestion. The active stroke of the pump is the opening of the oesophageal lumen by the contraction of radially arranged sarcomeres, which insert on to its cuticular lining, which typically has a triradiate lumen. The apices of the radii are folds which can act as spring-loaded hinges, allowing the lumen to open to a triangular cross-section, when muscles contract, and closing the lumen elastically on relaxation, as has been described in the median bulb of *Ditylenchus dipsaci* (Seymour 1977). In some nematodes, a click mechanism may operate with there being two stable configurations, open and closed. In some Dorylaimida complex cuticular plates may allow two bistable states (Bennet-Clark 1976).

The oesophageal pump must work against the hydrostatic pressure of the pseudocoelom, which will increase as fluid is pumped into the alimentary canal, until relieved by defecation. Repeated ingestion of fluid into the anterior oesophagus, before the posterior oesophagus empties into the intestine, will reduce the difference in pressure between the oesophagus and intestine. This is observable in many Rhabditida. Doncaster (1962) and Mapes (1965*b*) have analysed pumping in Rhabditida by cine photomicrography, the action being too rapid to easily observe by eye.

The engineering principles underlying the operation of the oesophageal pump have been discussed by Roggen (1973) and Bennet-Clark (1976), who have attempted to apply their models numerically to *Ascaris*, using physical data from other biological systems. In its simplest form the oesophagus can be treated as a hollow cylinder, with radially arranged muscles in the wall of the cylinder. Roggen suggests that the inner and outer bounding membranes are flexible, the cylindrical shape of the oesophagus depending on the hydrostatic pressure within the oesophageal tissues. A triradiate lumen, it is argued, minimizes the stresses within the bounding membranes while ensuring a stable configuration on closing. In the simple model, the configuration allows for all the muscles in the same cross-section to contract at the same rate to achieve the same proportional shortening.

Bennet-Clark's model for a cylindrical oesophageal pump takes into account the elastic stiffness of the cuticular walls of the oesophageal lumen. Contraction of the radial muscles will reduce the diameter and increase the length of the pump, since its volume will remain constant so long as the lumen is closed, and it is a mechanical principle that when the pressure in a hollow cylinder is raised, as it must be, then stresses in its walls are twice the axial or lengthwise stresses. When the internal pressure equals that of the hydrostatic pressure in the pseudocoel the situation changes. Continued contraction will deform the cuticle lining, the lumen will open and the pump will shorten while increasing in diameter. A three-rayed lumen will open to a trefoil cross-section. Calculations suggest that these relationships can reasonably be applied to *Ascaris* oesophagus.

The oesophagus contains an intrinsic nervous system which has been described in great detail in *C. elegans* (Albertson and Thomson 1976). It contains 20 neurones, including motor neurones, interneurones, and neurosecretory cells. Sensory input comes from proprioreceptive endings within the oesophagus, from the inner labial papillae, which are probably tactile and chemosensory, and from two neurones from the nerve ring (see Sections 2.8 and 2.11). Thus the oesophagus operates as a semi-autonomous system, receiving external and internal sensory inputs.

The interplay of locomotion and feeding have been described in *C. elegans* by Croll (1975a). In the absence of food, forward locomotion is periodically interrupted by brief reversals of the body waves, which may also be stimulated by contact with an object. Croll suggests that locomotion is controlled by two antagonistic intrinsic 'generators' initiating forward and backward body waves, respectively. In the presence of food, oesophageal pumping is associated with inhibition of locomotory body waves, though rapid side-to-side movements of the head 30–50 times a minute persist. Oesophageal pumping continues at over 90 cycles per minute for periods of 2–5 minutes.

The development of simple videotape cameras for recording microscopy has made such observations simpler to make (Seymour, Minter, and Doncaster 1978). When *Panagrellus silusiae* feeds on a suspension of carmine particles, rapid bursts of pumping by the anterior oesophagus (the corpus), at two or three times a second, in bursts lasting several seconds, draw droplets into the oesophagus while the bulb and isthmus remain closed (Mapes 1965b). The droplets coalesce at the hind end of the corpus, and are passed through the isthmus and posterior bulb into the intestine by their less frequent waves of contraction. The volume of the body increases until it has increased about 10 per cent, when it is relieved by defecation, at intervals of several minutes. *Rhabditis axei* feeds in a similar manner. The operation of the postcorpus muscles and valves in *P. silusiae* and *Rhabditis oxycerca* are described by Mapes (1965b) and Doncaster (1962).

The feeding of Tylenchida involves very rapid thrusting of the stylet; many times a second. Feeding has been described in *Ditylenchus* by Doncaster (1966, 1976), and Seymour, Minter, and Doncaster (1978). The metacorpus pumps with about six pulsations per second for varying periods of time, with stylet thrusting about four times a second. Two atypical methods of feeding in fungal-feeding Tylenchida have been described. In *Neotylenchus linfordi* forward and backward movements of the intestine, alternately compressing and expanding its thin-walled anterior lumen, act as a pump, while the soft base of the oesophagus may serve as a valve (Hechler 1962a). In *Hexatylus viviparus* the body has lower hydrostatic pressure than fungal hyphae on which it feeds to force food into the intestine (Doncaster and Seymour 1975). Perhaps correlated with this is frequent defaecation through the operation of a rectal pump, which may lower the hydrostatic pressure of body fluids (Seymour 1975a). Defecation has been described in several plant-feeding Tylenchida (Doncaster 1966), where the elimination of excess water ingested with the food is important.

3.4. Neurotransmission

In discussing neurotransmission it may be helpful to recall that body waves, head oscillations, and oesophageal pumping are largely under independent nervous control, and this is probably also true of copulation and oviposition (Croll and Wright 1976). Moreover, locomotion may be controlled by two mutually inhibiting 'generators' initiating respectively forward and backward body waves (Croll 1975a) (see Fig. 3.4). Most work has been done on neurotransmission in the larger animal parasites, especially *Ascaris suum*, in which pharmacological and electrophysiological methods can be combined (reviewed by Jarman 1976; Johnson and Stretton 1980). With free-living nematodes, because of their small size, conclusions must be drawn from the effects of externally applied drugs, microscopy and microspectrophotometry, though it has proved possible to implant electrodes (Tartar *et al.* 1977). The electron microscope shows several kinds of chemical synapses, typically making contact *en passant* between cells, and numerous gap junctions. There are many gap junctions

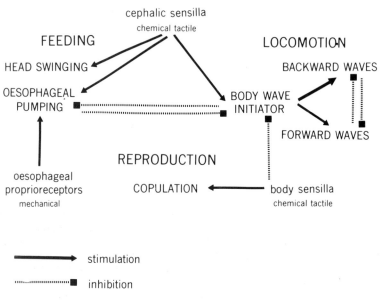

Fig. 3.4.   Schematic interpretation of neuronal co-ordination of feeding, locomotion, and, in the male, of copulation. These activities tend to be mutually inhibitory.

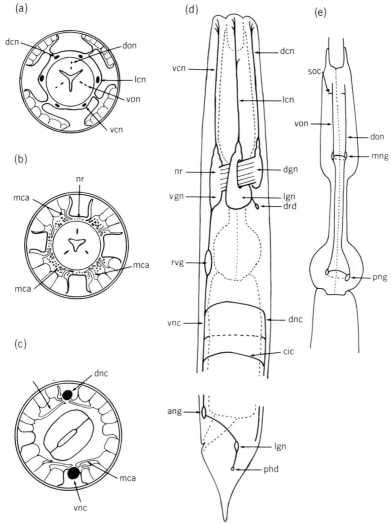

FIG. 3.5.   Diagrammatic representations of the nervous system of a nematode. (a)
Section through head; (b) section through nerve ring; (c) section through intestinal
region; (d) lateral aspect, showing nerves and ganglia on left side, those on right
side omitted; (e) oesophageal nervous system. don, dorsal (perradial) oesophageal
nerves; dcn, subdorsal cephalic nerve; lcn, lateral (amphidial) cephalic nerve; von,
ventrosublateral (interradial) oesophageal nerves; vcn, subventral cephalic nerve,
nr, nerve ring; mca, muscle cell arm; dnc, dorsal motor nerve; vnc, ventral motor
nerve; vgn, ventral ganglion; rug, retrovesicular ganglion; dgn, dorsal ganglion; lgn,
lateral ganglion; drd, deirid; cic, circumintestinal commissure; ang, anal ganglion;
lbn, lumbar ganglion; phd, phasmid; soc, nerve from somatic nervous system enters
oesophageal nervous system; mng, metacorpal nerve ganglion; png, postcorpus
nerve ganglion.

between the body-wall muscles and it has been generally assumed that electrical coupling through gap junctions allows co-ordination of intrinsic myogenic waves of contraction. The nervous system is shown diagrammatically in Fig. 3.5.

The evidence from larger parasites points to acetylcholine as the excitatory neurotransmitter (or possibly propionylcholine), associated with synaptic acetylcholine esterase (De Bell 1965; Jarman 1976; Johnson and Stretton 1980). The most likely inhibitor of somatic muscles is γ-aminobutyric acid, but the evidence so far is not convincing (Johnson and Stretton 1980). The importance of acetylcholine in soil nematodes is supported by the effectiveness of acetylcholine esterase inhibitors as nematicides. Acetylcholine may be important in sensory receptors and the central nervous system of Dorylaimida and Tylenchida, though the histochemical tests used do not always distinguish between acetylcholine and other esterases (Rhode 1960, Deubert and Rhode 1971). An esterase is secreted by the amphidial gland of *Meloidogyne javanica* and *M. hapla* (Bird 1966) and acetylcholine esterase was reported from the nervous system of two Dorylaimida by Lee (1964), and from the stylet region of one of them, *Actinolaimus hintoni*, a predator, where it may possibly paralyse its prey. Acetylcholine esterase is also secreted by the amphidial gland of a number of animal parasites (reviewed by McLaren 1976*b*), but its function is uncertain.

Formaldehyde-induced fluorescence has been used to detect another class of trnsmitters, catecholamines. Wright and Awan (1978) identified dopamine in the ganglia surrounding the nerve ring of *Prionchulus punctatus* (Dorylaimida), *Panagrellus redivivus*, and *Aphelenchus avenae*. Its presence in the lateral nerves of *P. punctatus* suggested a role in somatic muscle control. In *C. elegans* dopaminergic nerves are associated with cephalic sensilla, dereids, and male tail, which are probably mechanoreceptors (Sulston, Dew, and Brenner 1975). Dopamine, norepinephrine, epinephrine, and 5-hydroxytryptamine have been reported from homogenates of *C. briggsae* (Kisiel, Deubert, and Zuckerman 1976).

When applied externally to *P. redivivus*, *A. avenae*, and *C. elegans* (Croll 1975*b*), serotonin (5-hydroxytryptamine), 5-hydroxytryptophan, dopamine, and epinephrine stimulate vulval and vaginal muscle in *A. avenae* and *C. elegans* and copulatory activity in the male of *P. redivivus*. Somatic activity was depressed. Norepinephrine and histamine were without effect, apart from histamine stimulation of stylet extensions in *A. avenae*. The most striking observation was the prolonged rhythmic contraction of the vaginal and vulva muscles for many hours in *C. elegans* under the influence of serotonin. Croll (1975*b*) postulates that the direct introduction by the male of indolealkylamines (such as serotonin) may stimulate the female reproductive tract. In *C. elegans* they also stimulate the oesophageal bulb. Cholinergic inhibitors when applied superficially were without obvious effects. Willett (1980) was unable to recover serotonin from *P. redivivus*.

In reviewing the role of neurotransmitters in nematodes, Willett (1980) considers that the demonstration of the enzymes necessary to synthesize catecholamines in *Caenorhabditis*, as well as the effect of catecholamines in stimulating adenylate cyclase activity in *P. redivivus*, point to their potential role as transmitters, though the evidence is not unequivocal. Willett speculates that the precursors (e.g. dopamine), rather than the transmitters themselves, may accumulate in neurones, and clearance, following release, may depend on *O*-methyltransferase rather than recapture and degradation by monamine oxidase at the presynaptic junction. This would imply catecholinergic transmission. In *P. redivivus* at least some part of adenylate cyclase activity is catecholamine-sensitive, presumably further stengthening the argument for catecholamine transmission, coupled to cAMP as intracellular second messenger.

3.5. Behaviour

Nematodes respond to a variety of external stimuli, mechanical, light, temperature, electrical, and chemical, by moving towards or away from the source of stimulus, though the mechanisms involved are often open to question. External stimuli are also involved in feeding and copulation, but most work has been done on the locomotory responses of nematodes. Many early observations on nematode behaviour are discussed in a book on the subject by Croll (1970), who has more recently reviewed behavioural co-ordination (Croll 1975c, 1976a) and sensory mechanisms (Croll 1977). Croll has done much to elucidate the basic mechanisms underlying locomotory behaviour.

The response to chemical stimuli has been the most widely studied, especially to sex pheromones which serve to bring the sexes together prior to copulation. The first report of sexual attraction was published by Greet (1964), who observed that both males and females of *Panagrolaimus rigidus* attracted one another in agar. Since then sexual attraction has been reported in many species, and Bone and Shorey (1978), in reviewing nematode sex pheromones, concluded that sex pheromones have been reported from 23 species, and with one exception from every species in which studies have been reported. In every case males are attracted to pheromones released by the female, but in some species the female may also be attracted by a male pheromone. Pheromones have been reported from free-living, plant-parasitic, and animal parasitic nematodes (see Table 3.3). Apparently sexual attraction has not been reported from any Adenophorea or any marine nematodes.

TABLE 3.3
Sexual attraction in free-living nematodes

Species	Male attracted to female	Female attracted to male	Reference
RHABDITIDA			
Panagrolaimus rigidus	+	+	Greet (1964)
Pelodera teres	+		Jones (1966)
Pelodera strongyloides	+	+	Stringfellow (1974)
Rhabditis pellio	+		Somers, Shorey, and Gaston (1977)
Curznema lambdiense	+	+	Ahmad and Jairajpura (1981a)
Cylindrocorpus longistoma	+		Chin and Taylor (1969a)
Cylindrocorpus curzii	+		Chin and Taylor (1969a)
Panagrellus silusiae	+		Cheng and Samoiloff (1971)
Panagrellus redivivus	+	+	Samoiloff, Balakanich, and Petrovich (1974); Duggal (1978a)
Cephalobus persegnis	+		Cheng and Samoiloff (1971)
Acrobeloides sp.	+		Jairajpuri and Azmi (1977)
Chiloplacus symmetricus	+		Ahmad and Jairajpuri (1980)
TYLENCHIDA			
Heterodera (10 species)	+		Green (1966); Green and Plumb (1970)

The pheromones, whose chemical nature is unknown, are soluble in aqueous media, and the pheromone from *Heterodera* may be partially volatile (Greet, Green, and Poulton 1968). Concentration gradients may have a very different form in soil from that in agar in which the behavioural response has usually been studied, depending on rates of emission, diffusion, and loss of activity, particularly if volatility plays a part (Green 1980). Green concludes that the pheromone is likely to be an organic molecule, and that the decay of the gradient is more likely to be due to biological degradation than physical degradation or absorption. In *H. schachtii* diffusion appears to be rate-limiting on agar since the gradient is maintained for 20 hours after removal of the emitting female.

In a typical experiment a number of males are placed on agar with a choice of either moving up a concentration gradient towards a source of attractant, for example, virgin females, or towards a control, which does not release any attractant. The response is essentially statistical, with the majority of males progressively aggregating in the vicinity of the source of the attraction (Chin and Taylor 1969a; Cheng and Samoiloff 1971; Duggal 1978a). Variations on this general approach have shown that the pheromone is species-specific (Chin and Taylor 1969a; Cheng and Samoiloff 1971), or attractive to only closely related species (Green and

Plumb 1970; Samoiloff, McNicholl, Cheng, and Balakanich 1973). In *Pelodera strongyloides* OH⁻ ions, liberated by both sexes, are attractive to males, but a pheromone is also present attracting females to males (Stringfellow 1974).

Hydroxyurea blocks the development of the gonads of *P. redivivus* (Cheng and Samoiloff 1971), but not the development of the vulva and male copulatory organs. Pre-treated females are unattractive to males, while males cultured in hydroxyurea, do not respond to the pheromone. However, hydroxyurea is without effect on mature adults and does not prevent copulation or the birth of progeny. It seems that the release of pheromone by females and responsiveness of males are correlated with gonad development. Virgin females of *P. redivivus* are attractive, but inseminated females are not, and Duggal (1978*b*) speculates that eggs in the uterus without spermatozoa in the spermatheca are necessary for attractiveness. Males are constantly responsive and attractive. Similarly, in *Curznema lambdiense* young virgin females are attractive to males, but not after copulation (Ahmad and Jairajpuri 1981*a*).

The stimuli for copulation are different from attraction and require contact between the two sexes (Greet 1964; Jones 1966; Cheng and Samoiloff 1971; Duggal 1978*b*). Treatment of female *P. redivivus* with pepsin inhibits copulation (Cheng and Samoiloff 1972), which has been described in this species by Duggal (1978*b*, *c*), and in *C. lambdiense* (Ahmad and Jairajpuri 1981*b*).

Carbon dioxide may be either an attractant or a repellant in some nematodes depending on its concentration. In *P. redivivus* dilutions between 1:50 000 and 1:200 000 in water are attractive, 1:400 000 are without effect, while 1:20 000 is repellent (Balan and Gerber 1972). It seems probable that CO_2 attracts some plant-feeding species to plant roots. Reversible chemical inhibition of CO_2 receptors was demonstrated in *Ditylenchus dipsaci* (Tylenchida) by Croll and Viglierchio (1969*a*).

Some of the most detailed investigations of chemical attraction have been made with *C. elegans*. Ward (1973) studied the behaviour of *C. elegans* in gradients of many different chemicals in sephadex or agarose and reported attraction to a number of different chemicals (see Table 3.4). When a chemical attractant did not interfere with the response to a gradient of a second attractant, Ward concluded there were probably separate receptors for each chemical, as for Na⁺, Cl⁻, and cAMP. Studies of mutants, in which orientation in gradients of chemical attractants is impaired, suggest that the amphids may be the chemosensory organs, with different neurones responding to different chemicals (Ward 1976). One mutation impairs orientation to Na⁺, Cl⁻, and cAMP, but not to bacterial products. However, there are difficulties in demonstrating, unequivocally, that the mutation impairs a specific receptor and not some other step in the pathway between receptor and response, and in recognizing when a single gene influences several responses.

Dusenbery (1973) invented an ingenious and quite different method of studying the behaviour of *C. elegans* while swimming, called counter-current separation. In this method two solutions of different density, one of which can contain an attractant or a repellant, flow past each other with minimal mixing. The nematodes are introduced into one of the solutions and the response to the chemical is judged from the proportions of nematodes recovered from each of the two solutions. NaCl is a strong attractant, from a threshold of 0.1 to 50 mM, at temperatures from 15° to 25 °C, and pH from 3.0 to 9.0. Other ions are less attractive. Generally *C. elegans* is attracted to OH^- and avoids H^+. Its response to CO_2 depends on the buffer used (Dusenbery 1974).

Surprisingly *C. elegans* is atttracted to low concentrations of pyridine (0.1 – 1 mM) (Dusenbery 1976a), repelled by D-tryptophan over a similar concentration range (Dusenbery 1975), but does not respond to L-tryptophan. Dusenbery (1976b) used countercurrent separation to study 16 mutant lines of *C. elegans*, carrying single-gene mutations (Dusenbery, Sheridan, and Russell 1975), which failed to respond normally to NaCl. Their response to other attractants and repellants was measured, and most

TABLE 3.4

Chemicals that attract Caenorhabditis elegans

Class	Attractant	Accumulation threshold (mM)
1) Cyclic nucleotides	cAMP	0.2
	cGMP	0.2
	Not AMP, Bu_2−cAMP	
2) Anions	Cl^-	2
	Br^-	20
	I^-	20
	Not CH_3COO^-, F^-	
3) Cations	Na^+	2
	Li^+	4
	K^+	4
	Mg^{2+}	15
	Not NH_4^+, $CH_3NH_3^+$	
4) Basic pH	OH^-	~0.001
Unclassified	lysine	~10
	histidine	~10
	cysteine	~10

The attractants are shown grouped into classes, according to competition experiments. The accumulation threshold is the concentration of attractant at the centre of the plate that causes 25 per cent of the population (twice the background) to accumulate. Because of their high threshold, the amino acids have not been extensively studied except to show that the amino acids, and not their counter-ions, are the attractants. Reproduced from Ward (1973).

failed to respond, but some responded normally to some other attractants and repellants, or showed a reversed response. Dusenbery discusses the possible genetic basis for the observed phenotypes and concludes that there are probably at least six different receptors. Male *C. elegans* behave like the hermaphrodites (Dusenbery 1976c). Later (Dusenbery 1980) constructed a hypothetical dendogram to show the simplest explanation of how various mutations might block the pathway between receptors and motor output.

More superficial observations have been reported on the attraction of marine nematodes to food sources (Meyers and Hopper 1966; Gerlach 1977). *Chromadora lorenzeni* is chemically attracted to seaweeds to which it swims (Jensen 1981). The effects of bacteria on the dispersal of *C. elegans* and locomotion of *P. redivivus* have been described (Andrew and Nicholas 1976; Pollock and Samoiloff 1976). *Acrobeloides* sp and *Mesodiplogaster lheritieri* are attracted to suitable bacterial food in culture (Anderson and Coleman 1981).

Some nematodes respond to small temperature gradients; for example, the plant-feeding Tylenchida *Ditylenchus dipsaci*, *Tylenchorhynchus claytoni* and *Pratylenchus penetrans*, were attracted to warm spots in agar which had been raised by 1° above their surroundings (El-Sherif and Mai 1969), while two Dorylaimida were unresponsive, i.e. *Trichodorus christiei* and *Xiphinema americanum*. *D. dipsaci* collects at a 'preferred' temperature when placed in moist sand (Wallace 1961), depending on the temperature to which it was previously exposed (Croll 1967b).

Very interesting experiments have been made with normal and mutant lines of *C. elegans* by Hedgecock and Russell (1975). On agar, or in aqueous slurries of sephadex beads, *C. elegans* shows a temperature preference when placed on a thermal gradient. On agar with a radial temperature gradient, *C. elegans* moves in circles at a 'preferred' temperature, described as isothermal tracking. This temperature is determined by the temperature to which the nematode has been previously exposed and can be re-set by several hours' exposure. The authors suggest this re-setable eccritic temperature may lessen the chance that nematodes will be driven into unfavourable situations by normal temperature changes in the soil. However, if *C. elegans* has been starved at 20° it no longer accumulates at 20° when placed on the gradient, but disperses over the gradient. Some of the mutant lines of *C. elegans*, which fail to track isothermally, are cryophilic others thermophylic, suggesting two opposing drives may be balanced in the normal individual. Some of these thermotaxis-deficient mutants also failed to respond to chemical attractants (using countercurrent separation).

Sensitivity to light has been studied in several nematodes, and is presumably important in aquatic nematodes possessing eye spots or ocelli (discussed in Section 2.11). These are commonly found in nematodes associated with algae (Wieser 1959c). *Chromadorina viridis* moves towards light in the visible wavelengths, but away from u.v. light (Croll 1976a). *Diplolaimella schneideri* (Chitwood and Murphy 1964) and *Oncholaimus vesicarius* move away from the light (Burr and Burr 1975). However, not all nematodes with 'eye spots' have photoreceptors (Croll *et al*. 1975) and nematodes without apparent photoreceptors may respond to light.

Responses to touch have been studied quantitatively in *C. elegans* (Croll and Smith 1970; Croll 1976b). Generally the nematode tends to respond to contact by moving away, and when contacted on the head, to back away and change its direction of movement. The exploration, selection, and feeding of plant-feeding Tylenchida has been filmed and studied by Doncaster and Seymour (1973). Nematodes may respond to an electrical potential difference, i.e. galvanotaxis, usually by moving towards the cathode (Caveness and Panzer 1960; Whittaker 1969; Croll 1967a), possibly responding to its direct effect on the neuromuscular system. With *Anguina tritici* (Tylenchida), a plant parasite, the orientation of the larvae in an electric or magnetic field depended on the salt solution used, and Sukul, Das, and Gosh (1975) suggest that this may be because orientation depends on the ampids responding to charged ions. The response in *C. elegans* may also be to ionic gradients (Sukul and Croll 1978).

An interesting question is the mechanism of attraction to, and repulsion from, the source of stimuli. It is possible to account for the observed responses to some external stimuli as kineses. These are responses which alter the rates of movement (orthokinesis) or frequency of turning (klinokinesis) without any directional response to the stimulus. Klinokinesis could account for the aggregation of moving nematodes in favourable conditions and their avoidance of harmful conditions. Non-directional responses to heat, light, or gradients of chemicals may be important. More interest has been shown in possible taxes, in which the direction of movement is dependent on the direction from which the stimulus is received. The presence of paired sense organs suggests directional responses (though a single receptor can be sufficient).

The presence of paired ocelli, partially shaded by pigment, suggests phototaxis, and Burr and Burr (1975) interpret the movement of *Oncholaimus vesicarius* away from the light as a negative phototaxis. They propose the moving nematode orientates so as to balance and minimize the light received by the ocelli. With chemotaxis, the difficulty is that the proposed sensilla are so close together that they would have to detect very small concentration gradients if simultaneous comparisons were made (tropotaxis). The undulating motion of the head suggests that comparisons might be made at the extreme head positions, involving comparisons at different times (klinotaxis), but the difficulty is that the suggested sensilla are lateral, while the movements are in a dorso-ventral plane. None the less, Ward (1973), from studies of mutants of *C. elegans* moving in a gradient of attractant, concluded that klinotaxis was involved. The most telling mutant was one with a bent head, which moved towards the source of attractant along a spiral; with the body at an angle to direct path, but the head along that direction. Ward concluded that the head determined the direction, forward motion was unnecessary for orientation, as was fine control of its movement, but the span of movement was important (Ward 1976). Dusenbery (1980) calculates that the dimensions of ampids do not preclude klinotaxis.

Green (1966) proposed a different mechanism in *Heterodera* males approaching the female. In this scheme, the nematode moves forward until it detects a fall in concentration, then the head oscillates and the nematode turns or reverses before proceeding. Such a mechanism is compatible with many observations on the locomotory patterns in nematodes discussed in Section 3.4. A computer model of males responding to the female was used to simulate the males approach by Green (1977), assuming the above system and continuous klinokinesis. With male *P. silusiae*, sensilla at the tip of the spicules are important for orientation to the female (Samoiloff, McNicholl, Cheng, and Balakanich 1973). A 'two-state model', in which normal forward motion is interrupted by pheromone stimulation has been proposed for *P. redivivus* by Samoiloff, Balakanich, and Petrovich (1974). In an 'activated' state, the nematode makes rapid frequent sweeping changes of direction until differences in the intensity of stimulation at head and tail are maximized, when normal forward motion resumes. Orientation has been discussed more fully by Croll (1970, 1976a, 1967a), Ward (1976), and Dusenbery (1980). Examples of the tracks of nematodes as males are attracted to females are illustrated by Jairajpuri and Azmi (1977) with *Acrobeloides* sp. The presence of attractants, such as NaCl, and repellants, like D-tryptophan, alter the frequency of forward and backward waves in *C. elegans* moving on agar, but it is difficult to relate these changes to models of orientation (Rutherford and Croll 1979).

3.6 Osmoregulation and ionic regulation

Nematodes have no unequivocal osmoregulatory organs, though the alimentary canal, hypodermis, and the 'excretory' system have been implicated in experiments with some nematodes, mostly the larger animal parasites. With free-living nematodes, changes in volume, wet weight, and salt concentration in response to changes in the external environment have been studied, but have serious limitations. They fail to distinguish between changes within cells, within the pseudocoelom, and within the gut. Some studies have used solutions of NaCl, without considering the importance in osmoregulation of a balanced ionic environment, especially of Na, K, and Ca (Wright and Newall 1976, 1980).

Osmoregulation is not likely to have the same significance for marine nematodes as for terrestrial and freshwater nematodes, because of the constancy of the salt composition in the open sea, but intertidal and estuarine species are exposed to changes in salinity. *Chromadorina germanica* and *Monhystera denticulator* from shallow water (Tietjen and Lee 1972, 1977a) reproduced most rapidly at the salinities of their natural habitat, 26‰, but could grow and reproduce at a reduced rate in salinities between 13 and 39‰. Temperature interacted with salinity in influencing growth and reproduction. More extreme salinities, 6.5 and 52‰, were lethal. *Rhabditis marina* (Tietjen, Lee, Rullman, Greengart, and Trompeter 1970) collected from the same area of New York State, but also known from a variety of coastal habitats in many parts of the world, showed much greater tolerance. It could reproduce in salinities between 0 and 80‰, with an optimum of 45–55‰. *Diplolaimella schneideri*, an estuarine species, also tolerated de-ionized water for several days and

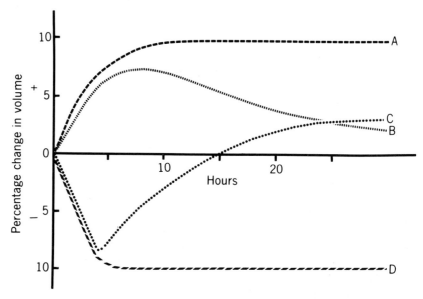

Fig. 3.6.   Osmotic volume regulation—a diagrammatic representation of how a hypothetical volume regulator would respond to hypertonic and hypotonic solutions by changes in its body volume: (A) in hypotonic solutions without compensatory changes in its internal body fluids; (B) in hypotonic solutions followed by the reduction in the tonicity of its body fluids; (C) in hypertonic solutions with the uptake of osmotically active substances; (D) in hypertonic solutions without internal changes in tonicity.

reproduced in salinities between 6‰ and 40‰ (Chitwood and Murphy 1964).

In a volume regulator, the nematode adjusts to changes in the external osmotic pressure by loss or gain of water, with consequent changes in volume (see Fig. 3.6). In hypotonic solutions the nematode will swell lowering the internal osmotic pressure and stretching the cuticle, muscles, and other organs until equilibrium is reached. Active regulation requires the secretion of water or solutes. In hypertonic solutions a volume regulator loses water and shrinks until it comes into osmotic equilibrium. Osmotic pressures can be balanced and the original volume recovered, i.e. regulated, by taking up solutes from the medium, but only so long as the body surface remains selectively permeable.

The large intertidal nematode *Deontostoma californicum*, from the hold-fasts of brown algae, was studied by Croll and Viglierchio (1969*b*), who used the length of the oesophagus as an index of volume changes. In hypotonic solutions of NaCl it behaved as a volume regulator, increasing by 15 per cent in distilled water and maintaining this volume for 24 hours. In hypertonic solutions it shrank but recovered over the next 3–4 hours within 5–10 per cent of its initial volume, suggesting hypertonic regulation. Ligaturing different regions of the body with silk thread suggested that water and ions passed through the body surface. Comparable studies on several Antarctic nematodes were reported by Viglierchio (1974). However, Wright and Newall (1976, 1980) are critical of this work because the authors used NaCl, or other pure salt solutions, instead of a physiologically balanced salt solution. They also argue that substituting length measurements for volume measurements may be misleading.

Newall studied two large coastal species of *Enoplus*, i.e. *E. brevis* an estuarine species and *E. communis* from rocky shores (summarized by Wright and Newall 1976, 1980). Both show impaired volume regulation in hypertonic solutions with Na as the only cation, but ionic regulations in solutions containing Na, K, and Ca. In the absence of both K and Ca, both species swell, even in concentrations isotonic with sea water. Both *E. communis* and *E. brevis* initially increase in volume when placed in diluted sea water, and decrease in more concentrated sea water. However, while *E. brevis*, the estuarine species, recovers its normal volume within 48 hours when the differences from normal sea water are not great (50–150 per cent), *E. communis*, a shore species, fails to do so, except from slightly hypertonic sea water (125 per cent).

Croll and Viglierchio (1969*b*) used hypertonic solutions to study the permeability of the body surface of *D. californicum* to different molecules. The nematode was collapsed or, 'exosmosed', in a 0.1 M solution through water loss, and then allowed to slowly recover its initial volume, or slightly exceed it, indicating that the molecules or ions had been taken up by the body restoring water loss and osmotic equilibrium. With salt solutions both anion and cation must enter the body so as to restore water loss, but by using different combinations of anions and cations, the permeability to each may be determined. The body was permeable to monovalent salt solutions, as well as Ca^{2+} and Mg^{2+}, but not to most polyvalent salts and non-electrolytes. However, strong monovalent salt solutions may well damage the tissues. The two species of *Deontostoma* studied by Viglierchio (1974) from the Antarctic gave similar results, but several species from freshwater genera failed to recover from exosmosis.

Terrestrial nematodes show varying degrees of hypotonic and hypertonic regulation. *Rhabditis terrestris*, taken from culture media, swells when placed in distilled water, but recovers its volume within about 30 minutes (Stephenson 1942). Recovery is prevented by 10 mM KCN. In balanced salt solutions of NaCl, KCl, CaCl, $NaHCO_3$ at pH 7.2 (Locke's solution) hypertonic to the culture medium, it shrinks and crumples, but recovers very slowly in only slightly hypertonic solutions, taking several days. The body wall changes in length more rapidly than the alimentary canal in both hypotonic and hypertonic solutions, and may rupture when swelling is rapid. The intestine apparently plays a major role in the elimination of water, and a change in the tonicity of the culture medium leads to some compensatory change in the tonicity of the internal body fluids. Viglierchio, Croll, and Gortz (1969) found that five species of terrestrial Tylenchida are more tolerant of osmotic change than *Rhabditis* sp Nematodes were transferred from salt solutions, or urea, to distilled water, sometimes causing *Rhabditis* to burst. Urea was less permeable than sodium while SO_4^{2-} was impermeable. The authors postulate that the fourth-stage larva of *Ditylenchus dipsaci*, which can tolerate desiccation, tolerates large internal changes in osmotic pressure without regulation, whereas *Rhabditis* tolerates only transient changes in its internal osmotic pressure, and regulates its internal pressure, unlike *D. dipsaci*. It will swell and burst when transferred from a strong solution to distilled water.

Osmotic changes are relatively slow in another terrestrial nematode *Panagrellus redivivus* studied by Myers (1966), taking many hours. The nematode responds to hypertonic solutions by losing water and taking up permeable salts, and to hypotonic solutions by swelling, followed by some loss of salts. *P. redivivus* appears to be relatively impermeable to glucose. The net uptake of water was delayed for several hours in hypotonic NaCl or glucose, perhaps by the elasticity of the cuticle or muscular contractions, but not in pure water. In balanced salt solutions internal Na^+ and Cl^- were maintained below their external concentration; K^+, Ca^{2+}, and Mg^{2+} above. Both K^+ and Ca^{2+} were taken up from the diet when available. Some Na^+, K^+, Cl^-, and Ca^{2+} are lost in distilled water. The ionic composition of *P. redivivus* and the tylenchid *Aphelenchus avenae* under

TABLE 3.5

Average inorganic ion content of P. redivivus *and* A. avenae *incubated in Fenwick's salt solution (FSS)*[1] *and distilled water (DW) for 24 hours at 22 °C*[2]

Nematode	Treatment[3]	mg element/g dry weight					
		Na	K	Ca	P	Mg	Cl
P. redivivus diet mainly bacteria	FSS	4.0	4.1	0.25	28.0	3.0	6.2
	DW	1.4	3.1	0.27	17.6	2.4	2.3
	FSS	4.6	4.9	0.26	25.0	5.0	3.8
	DW	1.7	2.7	0.24	18.0	5.6	0.5
diet mainly yeast	FSS	5.8	13.5	1.22	—	—	—
	DW	1.7	7.5	0.86	—	—	—
A. avenae diet fungus	FSS	3.2	4.8	0.4	—	—	—
	DW	0.9	4.5	0.5	—	—	—
	FSS	3.2	3.6	0.3	8.0	1.4	4.8
	FSS	9.6	2.7	0.4	12.9	0.9	8.2

Reproduced from Myers (1966).
[1] Fenwick's salt solution (FSS) in g/l: NaCl, 8.008; KCl, 0.201; $CaCl_2.2H_2O$, 0.265; $MgCl_2.6H_2O$, 0.203.
[2] The ranges of all Na, K, Ca, Cl, and P samples were within ± 5 per cent of the averages while the range of the Mg samples were about ± 10 per cent of the above reported averages.
[3] FSS and DW treatment in each experiment are related since analyses were performed on a single population of nematodes. Na, K, and Ca determinations in each experiment are related, whereas other ion analyses were made upon several separate populations of nematodes.

different cultural conditions are shown in Table 3.5, but these data do not show differences in the ionic composition of cells, extra-cellular fluids, and gut, nor which are osmotically active. Similar data for *Pelodera strongyloides* given by Scott and Whittaker (1970) are in mEq/l: $Na^+25.0$, $K^+2.0$, $CO_229.0$, $Cl^-19.0$; and mg/l: $Ca^{2+}6$ and P 20.

The permeability of the body surface of *A. avenae* to water and organic compounds when immersed in solutions of different osmotic pressure has been studied using tritiated water and ^{14}C-labelled compounds (Castro and Thomason 1973). The rates of exchange of water or organic compounds between the nematodes and the medium were calculated from the accumulation of the radioactive label with time. For rapidly penetrating substances, which reach a plateau over the course of an hour or so, the calculations are quite simple (see Table 3.6 which gives some examples from their data) but for slowly penetrating compounds more complicated calculations are required (Marks, Thomason, and Castro 1968). Water and acetone rapidly exchange between the nematode and the medium as do some organic nonpolar compounds, such as benzene, but most other polar

TABLE 3.6

Some examples of the rates of uptake and loss of water and organic compounds by A. avenae *from data given by Castro and Thomason (1973)*

Compound	k_{in} (min^{-1})	k_{out} (min^{-1})	$K_{equilibrium}$
Water	0.20	0.21	1.0
Acetone	0.21	0.22	1.0
Ethanol	0.025	0.031	0.8
Acetate (Na$^+$)	very small	—	—
Glycerol	very small	—	—
Glycine	very small	—	—
Glucose	0.00013	0.005	0.026
Benzine	0.5	0.05	10

Experiments with suspensions of 5×10^5 nematodes at 22 °C, using ^{14}C-labelled compounds and tritiated water.

compounds exchange very slowly. In hypertonic glucose, *A. avenae* comes into osmotic equilibrium by shrinking, in strong solutions (26 atm. O.P.) at a rate corresponding to the rate of water permeability. Wright and Newall (1980) have calculated the permeability constant for water for several nematodes using published data on experiments with deuterium oxide and tritium oxide. They give

  E. brevis and *E. communis* 3.4×10^{-4} cm s^{-1}
  A. avenae 1.3×10^{-6} cm s^{-1}
  C. briggsae about 2.0×10^{-6} cm s^{-1}

but consider these may be underestimates.

  Strong inorganic acids are toxic to *R. terrestris*, death taking several minutes in 0.2 N acids (Stephenson 1945), with survival inversely proportional to the logarithm of the normality, and influenced by the anion. The cuticle becomes sticky at pH below 3.5. *P. redivivus* is more resistant, tolerating pH as low as 3.0 (Ellenby and Smith 1966), and culturable at pH from 3.4 to 8.6 (Rothstein and Cook 1966), while *C. briggsae* can be cultured at pH 3.4–9 (my own observations), but is favoured by acid pH. *C. elegans* cultures suffer mortality when the pH rises above 6.5 (Vanfleteren 1976b). *T. aceti* shows extraordinary tolerance to changes in pH and to normally toxic compounds. It can be cultured in pH from 3.5 to 9, but grows best in the presence of acetic acid and at acidic pH (Rothstein and Cook 1966). It remains active in buffer solutions from pH 1.6 (KCl/HCl) to pH 11 (NaCl/glycine/NaOH). It also tolerates strong solutions of tannic acid, chromic acid, and strychnine (Peters 1928).

3.7 Respiration

Oxygen consumption by nematodes has interested nematologists from several different points of view. The biochemistry of cellular respiration will be considered in Chapter 5 and ecological energetics will be discussed in Chapter 8. Here I shall be concerned with the relationships between the partial pressure of oxygen (pO_2) and oxygen transport, oxygen consumption, and aerobic metabolism; these subjects have recently been reviewed by Atkinson (1980). Measurements of oxygen consumption can be used as a useful index of metabolic rate, but only when metabolism is strictly aerobic and the possible importance of anaerobic metabolism in free-living nematodes is controversial.

Oxygen consumption by large numbers of nematodes in a dense suspension can be measured by Warburg respirometry, or with an oxygen electrode, but for measurements on individual nematodes, Cartesian Diver microrespirometry is the appropriate method. Cartesian Diver respirometry has a long history in biology, but its application to nematodes has recently been improved and fully described by Klekowski (1971). The equipment is not expensive, but the technique requires considerable skill and practice. It can then give highly reproducible measurements of nematode respiration over periods of several hours. As usually employed, it measures the oxygen uptake of a non-feeding nematode, held in a small volume of water (about 2 µl) in a CO_2-free atmosphere at constant temperature, factors which need to be considered in interpreting the results. It can be adapted to measure the respiratory coefficient (RQ). Nematodes often move actively while in the Diver.

TABLE 3.7

Examples of the oxygen consumption by several nematodes from various habitats measured by Cartesian Diver respirometry at 20 °C

Species	Weight (µg)	Oxygen consumption		Habitat	Food	Ref.
		nl/individual /h	nl/µg wet w/h			
Panagrolaimus rigidus	0.59	1.47	3.02	Soil	bacteria	1
Cephalobus persegnis	0.49	1.15	2.72	ʺ	bacteria	1
Aphelenchus avenae	0.32	1.21	3.48	ʺ	fungi	1
Mononchus papillatus	5.22	4.30	0.79	ʺ	predator	1
Eudorylaimus ettersbergensis	0.30	0.29	0.97	ʺ	algae	1
Tobrilus gracilis	0.07–3.19	1.34–2.01	—	Lake benthos	unknown	2
Viscosia viscosia	0.84*	1.23	1.46*	estuarine mud	omnivor	3
Dichromadora cephalata	1.68*	0.99	0.59*	ʺ	algae?	3
Terschellingia communis	1.16*	0.54	0.47*	ʺ	unknown	3

* Assuming specific gravity 1.13.
References: 1. Klekowski *et al.* (1972); 2. Schiemer and Duncan (1974); 3. Warwick and Price (1979).

Pioneering studies were made by Overgaard Nielson (1949), but Klekowski, Wasilewska, and Paplinska (1972) have published much more extensive data on terrestrial nematodes and Warwick and Price (1979) on marine nematodes. For comparative purposes oxygen consumption is usually given as volume (at standard temperature and pressure) taken up per dry weight, per unit time, at constant temperature. With nematodes using the Diver these must be calculated from different parameters. The oxygen uptake by a single or sometimes several individuals is usually measured at 20 °C. The volume of the individuals can be calculated from measurements made later under the microscope after removal from the diver, using Andrassy's or Warwick and Price's formulae. Nematode weights must be calculated from their volumes and their specific gravity, while the dry weight may be taken as 25 per cent of the wet weight (see Section 2.1). Averages must be weighted means because of the relationship between oxygen uptake and volume (see below). The calculation of oxygen consumption at temperatures other than 20 °C requires a knowledge of the dependence of metabolism on temperature, often expressed as the Q_{10}, the coefficient for a difference of 10 °C (discussed in Section 3.8).

Metabolic rate R is a function of body size, usually measured by weight W. Within a group of morphologically similar animals

$$R = aW^b$$

There has been much discussion of the appropriate coefficient to use, but Klekowski *et al.* (1972) found from the regression of oxygen uptake on weight for 68 free-living, plant-parasitic, marine, freshwater, and terrestrial nematodes that the best fit was

$$R = 1.40 \pm 1.70W^{0.72 \pm 0.92} \text{ (nl O}_2\text{/individual h}^{-1}; W \text{ in µg).}$$

The metabolic rate per unit weight of tissue, or the intensity of respiration, is given by $R/W = aW^{(b-1)}$ or

$$R/W = 1.40W^{-0.28} \text{ (nl O}_2 \text{ µg}^{-1} \text{ h}^{-1}; W \text{ in µg).}$$

Analysing the data more closely it becomes apparent that there are differences in metabolic rates according to diet, predators and some plant parasites generally having higher rates than average, while some plant parasites were substantially lower. This also was true of the intensity of respiration. Klekowski *et al.* (1972) used Winberg's data (1971) (see Section 3.8) to recalculate values published by other workers at various temperatures for 20 °C. All their own measurements, on 22 species of soil nematodes, were made at 20 °C and are more reliable because they were all made under uniform experimental conditions. Respiration is, for example, influenced by the temperatures to which nematodes were exposed prior to the experiment. From their own observations

$$R = 1.7W^{0.67} \text{ (nl } O_2/\text{individual h}^{-1}; W \text{ in } \mu g)$$

and

$$R/W = 1.71W^{-0.33} \text{ } (R/W = \text{nl } O_2 \text{ } \mu g^{-1} \text{ h}^{-1}; W \text{ in } \mu g).$$

These figures show that respiration is a function of the surface area of the body, and that the smaller the nematode, the higher the 'intensity' of respiration. These generalizations hold true for many poikilothermic animals. Examples of the oxygen consumption by nematodes from various habitats are given in Table 3.7.

Oxygen consumption changes with growth. With *Panagrolaimus rigidus*, a bacteria-feeding soil nematode, Klekowski, Wasilewska, and Paplinska (1974), from whom the following data are taken, calculated a regression coefficient of respiration on weight of 0.67 of all stages, except the egg, where it was much lower.

Development stage	egg	L_1	L_2	L_3	L_4	Adult $\male$ & $\female$
Oxygen consumption per individual in nl O_2 h^{-1} at 20 °C	0.04	0.27	0.31	0.61	0.76	1.62
Intensity of respiration nl O_2 h^{-1} μg^{-1}	—	8.95	4.96	4.65	3.83	2.82

$$R = 2.02 \pm 1.28 \text{ } W^{0.64 \pm 0.12}$$

With a freshwater benthic nematode, *Tobrilus gracilis*, Schiemer and Duncan (1974) found the following equation fitted the data well for all the developmental stages

$$R = 0.522W^{0.693} \text{ } (n = 30, r = 0.898).$$

Warwick and Price (1979) carried out a similar study to Klekowski *et al.* (1972), calculating the intensity of respiration in 48 species of terrestrial, freshwater, and marine nematodes using their own and other published data. They took a value for $b = 0.75$. When the regression was examined with five species, 0.75 gave a good fit for four species, but a poor one for a fifth, a sluggish species, in which $b = 1.07$ gave the best fit.

Intensities can be compared by comparing the intercept log a' clustered round the mean value of -0.1, with range from $+0.188$ to -0.424. Generally it is possible to recognize four clusters of species: fast log $a' = 0.386–1.105$; medium-fast 0.057 to -0.081; medium-slow -0.091 to -0.199 and slow -0.207 to -1.081. A correlation between oxygen consumption and diet, as deduced from the structure of the buccal cavity had previously been suggested (Weiser and Kanwisher 1961; Teal and Wieser 1966). Warwick and Price concluded that omnivor/predators and non-selective deposit feeders have faster rates than epigrowth and non-selective deposit feeders (see Chapter 8 for terminology). Rather surprisingly, metabolic intensity was higher in larger species than small. Price and Warwick (1980) have also determined Q_{10} values for several species (Section 3.8).

Free-living nematodes are small enough for diffusion to meet their oxygen requirements when in water equilibriated with atmospheric air. Atkinson (1980) has considered many of the factors influencing the supply of oxygen to nematode tissues, applying diffusion equations from Hill (1929). As with most mathematical analyses of biological problems, some assumptions and approximations must be made, but it then becomes apparent that the body dimension and the partial pressure of oxygen in the environment are important. For example, by treating the body as a long cylinder, one can calculate the limiting body radius, r_0, which will permit oxygen to reach the axis of the body

$$r_0 = \sqrt{4ky_0/a},$$

where k is the diffusion coefficient of oxygen in tissue, y_0 the external partial pressure, and a the rate of oxygen removal by metabolism. It is apparent that r_0 is proportional to the square root of the partial pressure.

Atkinson's study shows that for nematodes with a body radius of 50 μm, a fairly typical nematode, diffusion is likely to be sufficient with partial pressures of oxygen as low as one-tenth that of air (15 mm Hg). Strictly, pO_2 should in future be expressed in SI units as newtons m^{-2}, but in the past most workers have expressed their results in mm Hg (1 mm Hg = 133.3 Nm^{-2}). In larger animal parasites, from oxygen-poor environments, diffusion alone will not supply the deeper tissues with oxygen, unless facilitated by body movements or haemoglobin.

The effect of reducing the pO_2 on the rate of respiration depends on the capacity of an organism to regulate its metabolic rate through oxygen binding and oxygen transport by respiratory pigments. Conformers use oxygen at a rate proportional to the pO_2 while regulators instead maintain a steady consumption with falling pO_2, until a critical value is reached, when it falls very rapidly. The latter relationship suggests respiration mediated by cytochromes. In laboratory experiments, C. briggsae (Bryant, Nicholas, and Jantunen 1967), Pelodera strongyloides, and C. elegans (Bair 1955) displayed critical values in aqueous solutions equilibriated with atmospheres containing 5, 8 and 16 per cent O_2, respectively. Anderson and Dusenbery (1977) checked Bair's value for C. elegans by a more sensitive method, using an O_2 electrode, confirming that it behaves as a regulator, but reporting a critical value of 27.4 ± 4.9 mm Hg (equivalent to an atmosphere of 3.6 per cent O_2).

Two intertidal species of Enoplus (Enoplidae) differ in their sensitivity to lack of oxygen, which is correlated with the availability of oxygen in their habitat. *Enoplus communis*, found amongst seaweeds or on *Mytilus* shells, becomes quiescent in the absence of oxygen, but is more tolerant of low levels of oxygen (Atkinson 1973a, b). Both species have haemoglobin in the muscles of the body wall, identified by microspectrophotometry from its characteristic oxygenated and deoxygenated absorption spectra (Atkinson 1977). However, only *E. brevis* has measurable amounts of haemoglobin in its oesophagus. Oxygen is progressively released from oxyhaemoglobin in the oesophagus as the pO_2 is reduced from 20 to 5 mm Hg. Electrical stimulation of muscular activity also causes deoxygenation. There is probably too little haemoglobin to serve as a useful store of oxygen, but it probably facilitates the diffusion of oxygen into the oesophagus under conditions of low pO_2 (Atkinson 1975, 1980). Experiments have shown that the feeding of *E. brevis* is less affected by low pO_2 (between 8 and 40 mm Hg) than that of *E. communis*.

Activity may be reduced in oxygen deficiency with the nematode becoming quiescent under anaerobic conditions. The capacity of soil nematodes to tolerate anaerobiosis in a quiescent or cryptobiotic state varies greatly. *Caenorhabditis* sp. are rather sensitive (Cooper and Van Gundy 1970), whereas *Aphelenchus avenae* enters a cryptobiotic state. An atmosphere of 5 per cent O_2 seems to be a critical lower level. Under laboratory conditions, with mixtures of oxygen in nitrogen, the growth and reproduction of *C. briggsae* was apparently normal with 4.4 per cent O_2, reduced at 3.7 per cent O_2, and inhibited at 0.17 per cent (Nicholas and Jantunen 1966). It was retarded slightly by the addition of 5.2 per cent CO_2 (in solutions buffered between pH 5.5 and 8.0). *C. briggsae* survived for 24 hours, but not 48 hours, in oxygen-free nitrogen or hydrogen, while the eggs survived three days, but not six (Nicholas and Jantunen 1964). *Acrobeloides beutschlii* recovered after eight days in an atmosphere of N_2 (Nicholas 1962). *Pelodera punctata* from sewage sludge survived 14 days anaerobically (in the presence of bacteria) and tolerated high levels of CO_2 (Abrams and Mitchell 1978). *Mesodiplogaster lheritieri* swallows air at the surface of culture media (Klingler and Kunz 1974), thereby escaping oxygen deficiency.

Nematodes from swamps and the benthos of lakes may be more often exposed to anaerobic conditions and tolerate the conditions better. In several uncontrolled laboratory experiments *Dorylaimus* sp. recovered from tropical papyrus swamps survived 86 days of oxygen deficiency in a quiescent state, recovering when oxygen was admitted, but in other experiments quickly succumbed (Banage 1966). *Eudorylaimus andrassy* from the benthos of Lake Tiberias (Israel) survived six months' anaerobiosis in mud in a laboratory (Por and Masry 1968).

Marine and estuarine sediments are often anaerobic about one centimetre beneath the surface through microbial action, coloured black by ferrous sulphide, and rich in H_2S (Fenchel and Riedl 1970; Ott and Schiemer 1973). However, Reise and Ax (1979) do not believe that a diverse population of nematodes or other meiofauna inhabits the anaerobic sediments, but that the fauna clusters around oxygenated worm burrows; they base their conclusions on studies of the fauna of intertidal mud flats.

Experiments have shown that nematodes from the anaerobic region survive for many days in the laboratory in media flushed with oxygen-free nitrogen or hydrogen in sealed containers (Moore 1931; Wieser and Kanwisher 1961; Ott and Schiemer 1973). Ott and Schiemer (1973) found nematodes from the anaerobic zone of an intertidal sandy beach survived for 2–12 days, in sea water equilibriated with oxygen-free nitrogen, while those from the surface layers survived for only a few hours. These authors concluded from a re-examination of data published by Wieser and Kanwisher (1961) and Teal and Wieser (1966), that, when allowance was made for differences in weight, the respiration rate (intensity of respiration) was lower in nematodes from marine and limnic sediments than in those from more aerobic sediments. Perhaps low rates of aerobic respiration are coupled with concomitant anaerobic metabolism. However, the intensity was also correlated with feeding habits. Whether any marine nematodes respire anaerobically, as some parasitic nematodes have been shown to do, is unknown, because appropriate techniques have yet to be developed of comparable sensitivity to the Cartesian Diver microrespirometry for anaerobic respiration, and suitable methods of axenic culture are required to provide material for a study of their biochemical pathways.

The respiratory quotient, RQ, the ratio of CO_2 evolved to O_2 consumed, is indicative of the food reserve being catabolized (0.95–1.0 for carbohydrates; 0.74 for lipid plus perhaps some protein). Glycogen and fat are utilized as reserves in *Caenorhabditis* sp and *A. avenae*, with an RQ of 0.94–0.96 when well fed, falling to very low values in starvation, close to 0.4 (Cooper and Van Gundy 1970). Such low values were also found in *Panagrellus redivivus* by Santmyer (1956), perhaps indicating CO_2 consumption in the glyoxylate or other cycles, calling into question the value of RQ in nematodes. Values close to 9.74 have been reported in *P. redivivus* (Hammen 1967) and another bacteria-feeding nematode *Pelodera chitwoodi* (Mercer and Cairns 1973).

3.8. Temperature

Where sufficient moisture is present some nematodes are to be found active at all temperatures between 0° and 47 °C. Survival outside these limits is in an inactive cryptobiotic state, which will be considered in the next section. The range of temperatures tolerated by particular species may however be quite narrow and must therefore play a significant part in limiting distribution. Species and geographical subspecies of *Rhabditis* from the tropics tolerate higher temperatures than those from temperate regions (Sudhaus 1980*a*). Nematodes are found in thermal springs (Hoeppli 1926; Meyl 1953*a*, *b*, *c*; Kahan 1969). Most species reported from hot springs are cosmopolitan species, which have been found in less extreme conditions elsewhere, including bacteria-feeders, algal-feeders, and predators; but some species and subspecific varieties seem to be restricted to thermal waters (Hoeppli 1926; Meyl 1953*a*, *b*). A temperature of 45–47 °C probably marks the upper limit for sustained inhabitation, though nematodes have been reported from higher temperatures. Behavioural response to temperature gradients, and behavioural acclimatization to temperature has been discussed (Section 3.5). The genetics of temperature adaptation will be discussed in Chapter 5.

Some observations have been made on tolerance of low temperatures above the freezing point in several species of nematodes by Lyons, Keith, and Thomason (1975). They looked for discontinuities in the typical linear relation between oxygen uptake and temperature (Arrhenius plots), matching the temperature of the discontinuity with the temperature at which phase changes were found in phospholipids by electron spin resonance. *Meloidogyne javanica* and *C. elegans*, which do not tolerate temperatures below 10 °C, show a sharp change in respiration and a phase change in their phospholipids at this temperature. *M. hapla* and *Anguina tritici* are more tolerant of chilling, show no such change in respiration, and no phase change. *A. avenae* also tolerates a wide temperature range, but does show a phase change at 20 °C, or a change in oxygen uptake. The authors suggest that phase changes in phospholipids alter the activity of membrane-bound enzymes, and that these are not tolerated by the more temperature-sensitive species.

The most obvious effects of temperature are its effects on growth rate, generation time, and fecundity. Equations expressing growth and fecundity as functions of temperature are required for mathematical modelling of nematode populations, and several attempts have been made to describe the relationship mathematically. Jones (1977) used data published by Grootaert and Maertens (1976) on *Mononchus aquaticus*, a nematode predator, to derive the empirical equation

$$R = (A + BT + CT^2) / (1 + DT + T^2),$$

where R is the rate of development (1000/gen. time in days), T is the temperature in °C, and A to D are coefficients. Later Greet (1978) used the same equation for the bacteria-feeding soil nematode *Panagrolaimus rigidus*.

Coefficient	M. aquaticus	P. rigidus
A	− 6.94713	− 24.9778
B	1.67753	5.81649
C	− 0.04564	− 0.16613
D	− 0.05556	− 0.05972
E	0.00078722	0.00092152
Threshold for development	5–10°	7°
Optimum and generation time	28° 14–15 days	20° 5 days
Optimum for egg production	22–8°	20°
Egg production at optimum	22° (about 90)	15° (about 20)

A different approach was used by Hiep, Smol, and Absillis (1978) with *Oncholaimus oxyuris*, a brackish water nematode predator with only one generation a year. Their interest lay primarily in its generation time and reproductive potential in its natural habitat, rather than in the effects of temperature *per se*, though both depend on temperature. They drew up a life-table from laboratory data (Poole 1974) and found a linear relationship between temperature and the intrinsic rate of population growth. The relationship between the intrinsic daily rate of population increase, $r.d^{-1}$, and temperature is quite different for laboratory populations of another estuarine nematode, *Diplolaimelloides brucei* (Warwick 1981), primarily because of the sharp increase in fecundity with temperature between 10 and 25 °C. The regression of r on temperature was given an approximate solution using the equation

$$r = -0.967 + 0.0134 \, T.$$

Warwick used an equation of the form

$$\text{Generation time} = aT^b$$

to describe the effects of temperature on growth rate. b for *D. brucei* was within the range calculated by Warwick for a number of marine species from published data ($b = -1.915$; days °C).

Data on the effects of temperature on growth rate and reproduction can be found in many of the studies of nematodes in culture referred to in Chapter 4. Examples for terrestrial nematodes are to be found in Pillai and Taylor (1968), Grootaert (1976), Yeates (1970), and Grootaert and Jaques (1979) and for marine nematodes in Tietjen and Lee (1972), Gerlach and Schrage (1971, 1972), Tietjen *et al.* (1970), Hopper, Fell, and Cefalu (1973), and Warwick (1981).

The effects of temperature on respiration are often expressed by the Q_{10}. A theoretical basis for its use has been given by Winberg (1971), together with empirical values derived from Krogh's (1916) normal curve

Temperature (°C)	5–10	10–15	15–20	20–5	25–30
Q_{10}	3.5	2.9	2.5	2.3	2.2

For *C. elegans* the Q_{10} (16–25°) for development is 2.1 (Byerly, Cassada, and Russell 1976). Price and Warwick (1980) studied the effects of temperature on the respiration of estuarine nematodes by Cartesian Diver microrespirometry (more fully discussed in Section 3.7). The respiration rate Rc (O_2 uptake) per unit body volume was plotted over the range to which the animals were exposed in nature (5°–20 °C) in the form

$$\log_{10} \text{Rc} = c + \text{d.T.}$$

Q_{10} values were calculated: i.e. $\log Q_{10} = 10d$. For *Trefusia schiemeri* Q_{10} was about 2; for *Sphaerolaimus hirsutus*, 1.17 (but only within the normal environmental temperature range). The authors discuss these values in the wider context of meiofaunal dietary habits. With *Diplolaimelloides brucei* from the same habitat the Q_{10} (5–25 °C) = 3.94 (Warwick 1981).

Santmyer (1956) measured the oxygen consumption of *Panagrellus redivivus* at different temperatures, finding an approximately linear relationship between 10 and 35 °C. Temperature also affected locomotion, the rate of propagation of body waves rapidly increasing to a plateau between 25° and 35 °C then falling rapidly. High temperatures change the sex ratio in *P. redivivus* from 1:1 to a high frequency of males (Hansen and Cryan 1966b), but do not change the sex ratio of *D. brucei*, steady at 2:1 in favour of males (Warwick 1981).

3.9. Quiescence, diapause, and cryptobiosis

Many nematodes pass through stages of inactivity and reduced metabolism as part of their life-cycle. Such stages may be induced by environmental conditions or inherent in their development. Inactive stages are most obvious in the life-cycles of animal and plant parasites, where these conserve the energies of infective stages and facilitate their dispersal. Many terrestrial free-living nematodes survive extremes of cold, desiccation, and heat in an inactive state, enabling them to inhabit places with cold winters and deserts. In aquatic nematodes, anoxia and osmotic stress can inhibit activity and threaten survival. It is necessary to distinguish different kinds of inactivity. Van Gundy (1965), reviewing the subject, distinguished quiescence, diapause, and cryptobiosis. Inactivity with reduced metabolism was termed quiescence. When nematodes enter a quiescent state, whether from the effects of environmental change or from an intrinsic developmental process, but do not resume metabolism and activity under favourable conditions until receiving a specific extrinsic or intrinsic stimulus, the condition is called diapause. Survival in conditions of complete metabolic arrest is called cryptobiosis (anabiosis, abiosis). The complete absence of metabolism may be impossible to prove directly, but may be presumed in nematodes which tolerate very low temperatures, high degrees of desiccation, or high temperatures with total inactivity.

Evans and Perry (1976) in reviewing the subject used a different terminology, using dormancy to cover all forms of metabolic inactivity, distinguishing between diapause and quiescence, with the latter including various categories of cryptobiosis, such as anoxybiosis and cryobiosis, depending on the environmental stress. However, once in a cryptobiotic state, nematodes tolerate a variety of stresses including heat, cold, and desiccation, the common feature being a change in the state of hydration of their tissues.

Diapause is found in the eggs of specialized plant-root parasites, ensuring that some viable eggs persist in the soil for long periods in the absence of host plants. It has been most thoroughly investigated in Heteroderidae (reviewed by Clarke and Perry 1977). A proportion of the eggs containing larval nematodes fail to hatch under favourable conditions, entering diapause, which may be 'broken' in some species by low temperature (e.g. Ogunfowora and Evans 1977), or in others by an intrinsic metabolic change (e.g. Guiran 1979; Meagher 1975).

Many free-living Rhabditida produce 'resting' dauer larvae. In *C. elegans* quiescent and resistant dauer larvae are produced when food is scarce, which resume development in the presence of food (see Chapter 6). In other species the dauer larva resists drying, serving to disperse the nematode on the bodies of insects. Many larval plant parasites can tolerate conditions of freezing and desiccation in a cryptobiotic state for long periods, facilitating their dispersal. The larvae of *Ditylenchus dipsaci*, for example, have been revived from dry plant material after 23 years (Fielding 1951) and also after lyophilization at $-80\,°C$ (Bosher and McKeen 1954).

Survival when dried depends on the conditions in which the nematode is desiccated. In *Aphelenchus avenae*, the proportion of nematodes surviving is strongly influenced by the rate of dehydration (Crowe and Madin 1974, 1975). When desiccated at relative humidities of 97 per cent for 72 hours before being transferred to dry air, many survive on rehydration, but more dehydration at lower humidities leads to much reduced rates of recovery on rehydration. Survival is also enhanced by exposing nematodes to humid air (at 97 per cent RH) for some time before immersing them in water. Entry into cryptobiosis, or more specifically anhydrobiosis, is an active process, which has been studied in detail in *A. avenae*, using nematodes desiccated at 97 per cent RH for 72 hours. Slow dehydration is important in other cryptobiotic nematodes, which may be facilitated by coiling

(Demeure, Freckman, and van Gundy (1979*b*). Slow drying is important in *Rotylenchus robustus* for survival, perhaps associated with an orderly shrinkage of the cuticle (Rossner and Perry 1975). Slow dehydration and adequate nutrition enhance the survival of dehydrated *Aphelenchoides besseyi* (Huang and Huang 1974).

Genera from several orders of nematodes were extracted from Mojave desert soils in a tightly coiled state by Freckman, Kaplan, and van Gundy (1977) in 1.25 M sucrose, which became active when rehydrated. Nematodes from desert soil coiled before the soils became very dry. Experiments with *Acrobeloides* sp *A. avenae* and *Scutellonema brachyurum* (Demeure, Freckman, and van Gundy 1979*a*) desiccated under controlled conditions in the laboratory, suggest that coiling is induced by forces exerted on the nematode by soil moisture, becoming effective when the water layer was 6–9 monomolecular layers deep. Other nematodes aggregate forming a 'wool' which slows water loss and increases survival when dehydrated (Ellenby 1969).

As *A. avenae* enters anhydrobiosis, glycogen and lipid reserves are reduced, while glycerol increases from nondetectable levels to 45–55 μg mg^{-1} dry weight, and trehalose from 20 to 92–125 μg mg^{-1} dry weight (Madin and Crowe 1975; Crowe, Madin, and Loomis 1977). On rehydration, the levels of glycerol and trehalose fall again over several hours, while glycogen, water, and oxygen consumption increase. A water level of 2.5 μg mg^{-1} dry weight seems a critical value below which survival is correlated with glycerol and trehalose content. The importance of glycerol is interesting because of the large body of biomedical research on the use of glycerol to preserve vertebrate tissues when frozen and then thawed. One hypothesis is that glycerol replaces water molecules hydrogen-bonded to the polar groups of phospholipids and proteins, stabilizing their structure when chilled or dehydrated by freezing. Other hypotheses have been that glycerol protects phospholipids from peroxidation when dry (Crowe and Madin 1975), and that glycerol prevents a phase change in cell-membrane lipids increasing permeability. The latter was tested experimentally with *A. avenae* and it was found that on rehydration, anhydrobiotic nematodes were less leaky to ions and primary amines than rapidly dried nematodes. The correlation between glycerol synthesis and reduced leakiness to amines, suggested glycerol might be responsible by H-bonding to the polar groups of phospholipids (Crowe, O'Dell, and Armstrong 1979). Womersley (1981) reviewed the subject of anhydrobiosis and concluded that both the role and the importance of glycerol for anhydrobiosis was doubtful.

Womersley and Smith's (1981) work throws doubt on the importance of glycerol but emphasizes the importance of trehalose and inositol. Comparisons between the composition of two cryptobiotic species,

Anguina tritici and *Ditylenchus dipsaci*, when anhydrobiotic and when hydrated, with three intolerant species, *D. myceliophagus*, *Panagrellus redivivus*, and *Turbatrix aceti*, were very interesting (using gas chromatography to give composition as a percentage of dry weight). Trehalose was very high in both cryptobiotic species when desiccated but fell on rehydration, whereas it was not high in the other three species. Glycerol was higher in *T. aceti* and *P. redivivus*. Inositol was high in *A. tritici* when desiccated falling on rehydration. Womersley (1981) believes the importance of trehalose in anhydrobiosis may be that this non-reducing sugar does not react with amino acids and proteins when dry (browning). Studies of the carbohydrates and lipids of these same five species (Womersley, Thompson, and Smith 1982) showed that glycogen was stored by the sensitive species, but not by *A. tritici and D. dipsaci* which stored trehalose preferentially. No significant differences were found in lipids or phospholipids.

Trehalose is derived from glycogen in *A. avenae*, the regulatory enzyme being trehalose phosphate synthetase, controlled by the rising level of uridine diphosphate glucose during desiccation (Loomis, Madin, Crowe 1980; Loomis, O'Dell, and Crowe 1980).

A. tritici which forms galls on Gramminae, survives long periods in dry galls, at least 27 years (Fielding 1951). Electron microscopy shows changes in the cuticle and an orderly packing of the tissues when cryptobiotic (Bird and Buttrose 1974). The concentration of ATP (measured by luciferin luminescence) remains constant 1.0×10^5fg), rising to about twice this value on rehydration (Spurr 1976). Higher levels have been reported in *A. avenae*, i.e. 2.23 ± 0.33 µg mg^{-1} in the hydrated state, 4.59 ± 0.78 in the anhydrobiotic state, by Willett, Freckman, and Van Gundy (1978).

Cryptobiotic nematodes tolerate higher and lower temperatures than active individuals, as well as desiccation. *A. tritici* larvae survived brief exposure to 105 °C (2 minutes). Anhydrobiotic *Scutellonema cavenessi* from Senegal survived temperatures of 50 °C for several hours, whereas the active form did not (Demeure 1978). Survival in hot dry conditions may be important in this nematode. Two Dorylaimida recovered from dry mud by Lee (1964) possessed remarkably heat-stable esterases: active after six hours at 100 °C in dry nematodes but not after 10 minutes at 90 °C when wet.

Cryptobiosis may be involved in survival under adverse conditions other than extremes of temperature or desiccation. Cooper and Van Gundy (1970) consider that *A. avenae* enters a cryptobiotic state in oxygen deficiency, and Lees (1953) observed the survival of *P. silusiae* in hyptertonic glucose (4M) in an inactive state. However, it is difficult to be sure that metabolism is completely halted in these conditions and to distinguish cryptobiosis from quiescence.

Nematodes that do not enter cryptobiosis under natural conditions may experimentally be stored at very low temperatures. Hwang (1970) reported that *C. briggsae*, *P. redivivus*, *T. aceti*, and *Aphelenchoides sacchari* (=*A. rutgersi*) recovered after six months storage in a liquid N_2 refrigerator. Dimethyl sulphoxide (5–10 per cent) was added to aid survival, and the conditions of freezing and thawing were critical. Sekiya (1966) used glycerol (7 per cent) in a medium in which *Rhabditis elongata* was successfully stored for 120 days at -79 °C. Storage when frozen in a liquid N_2 refrigerator has now become widely used to maintain stocks of *C. elegans*, including many mutant lines. Brenner (1974) describes the method of storage using glycerol in the medium.

4. Culture, food, and nutrition

4.1. Introduction

MANY free-living nematodes are easily cultured in the laboratory for indefinite periods, and the short generation times, sometimes only a few days, together with their great fecundity, makes it possible to produce them in large numbers for experimental purposes. This has made free-living nematodes suitable for experimental work in many different branches of biology. Difficulties with their taxonomy can be avoided by using well known species, which can be easily exchanged through the post. The fact that genotypes of interest to experimental biologists can be stored in liquid N_2 refrigerators further enhances their use for experimental purposes. Of course, for many purposes, nothing so elaborate is required and cultures of locally collected species in simple culture media may suffice.

One objective of experimental biologists has been to establish pure cultures in the absence of micro-organisms in a chemically defined medium. Dougherty set out to establish axenic cultures of Rhabditidae in chemically defined media for genetic experiments in the late 1940s, a task which proved unexpectedly difficult and has only quite recently been achieved. Dougherty coined the terms axenic (free from strangers), monoxenic (with one other type of organism), and agnotobiotic or xenic (for cultures containing unknown kinds of other organisms). These terms have proved convenient, though I prefer xenic to Dougherty's agnotobiotic. The unknown associates are for practical purposes the microbial fauna. Many years of work in several laboratories chemically defining a suitable axenic culture medium has provided much information on nematode nutrition and biochemistry. However, for genetic experiments and much work in molecular and developmental biology, inspired by Nigon in France and by Brenner in England, monoxenic cultures have proved quite satisfactory, if not superior. For this work, largely with species of *Caenorhabditis*, agar cultures with the nematodes feeding on a well characterized strain of the bacterium *Escherichia coli* have been used.

4.2. Xenic cultures

Many nematodes can be cultured very simply in petri dishes in nutrient agar, as first reported by Metcalf (1903). Its clarity makes it ideal for observations of the living nematode, and, with variations in its stiffness and nutrient content, easily subcultured stable cultures of many different kinds

100

of nematodes can be established. Various standard microbiological agar media have been used, for example bacteriological nutrient agar, malt extract agar, and glucose tryptose agar, the aim being to favour the growth of the nematode and its microbial food over other organisms. The method has been so widely used that it will be impractical to attempt comprehensive lists of references. Generally, the technique is applicable to bacteria-feeding Rhabditidae, Diplogasteridae, Cylindrocorpidae, Cephalobidae, and Plectidae, fungal-feeding Tylenchida and Aphelenchida, and nematode-predatory Dorylaimida. As an example, Pillai and Taylor (1968) cultured about 15 different species from sewage in sucrose tryptose agar and failed with only three species. Some Diplogasteridae are opportunistic feeders which will feed on micro-organisms, or other nematodes when available. Pillai and Taylor (1968) found *Paroigolaimella bernensis*, *Fictor anchicoprophaga*, and *Butlerius* sp. fed on *Aphelenchus*. Examples of such Diplogasteridae are *Mononchoides potohikus* from sand dunes (Yeates 1969), *Mononchoides changi* from sewage (Goodrich, Hechler, and Taylor 1968), and *Butlerius degrissei* from soil (Grootaert and Jaques 1979).

The Dorylaimida includes nematodes with a wide variety of feeding habits, and even a single species may have a varied diet. *Dorylaimus ettersbergensis* in water agar cultures fed on blue and green algae, fungal conidia, ciliates, and other nematodes (Hollis 1957). A blue-green alga *Chroococcus* proved the best food. *Aporcelaimellus* sp. was cultured in agar on algae by Wood (1973*a*), who found that it required large spherical algae like *Haematococcus* to feed. A similar dorylaimid, *Labronema* sp., which like both the previous species feeds with an odontostyle, can be cultured on a diet of another nematode, *Panagrellus redivivus* (Grootaert and Wyss 1978). The culture of Mononchidae has interested several nematologists because of their possible role in controlling other nematodes. The Mononchidae, which have a cavernous buccal cavity with a lateral tooth (Fig. 4.1), feed voraciously on other nematodes, but also require bacteria in their diet. The earliest description of their culture and feeding habits was reported by Steiner and Heinly (1922) working with *Mononchus* (= *Clarkus*?) *papillata*. More extensive studies have been made on *Prionchulus punctatus* (Grootaert and Maertens 1976; Small and Evans 1981). These nematodes will feed on a variety of other nematodes, for example, *Panagrellus redivivus* and *Aphelenchus avenae*, which are easily cultured *en masse*, and soil extract agar provides a suitable base on which the necessary soil bacteria grow. Apin and Kilbertus (1981) have shown bacteria in the gut of several Mononchidae by electron microscopy. The aphelenchid *Seinura tenuicaudata* can similarly be cultured (Hechler 1963; Wood 1975).

Some aquatic nematodes cannot be cultured in agar, but will reproduce in water containing several wheat grains, or a little breakfast cereal, or

(a)

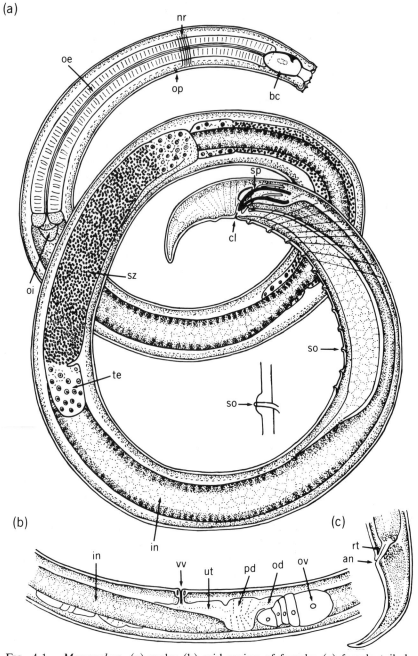

FIG. 4.1. *Mononchus*. (a) male; (b) mid region of female; (c) female tail. bc, buccal cavity; nr, nerve ring; op, excretory pore; oe, oesophagus; oi, oesophago-intestinal valve or cardia; sz, spermatozoa; te, testis; in, intestine; so, supplementary organ; sp, spicule; cl, cloaca; vv, vulva; ut, uterus; pd, pars dilator of oviduct; od, oviduct; ov, ovary; an, anus; rt, rectum.

other such food, though other organisms especially fungi and ciliates grow excessively. If the contaminants can be controlled, very dense cultures of some bacteria-feeding nematodes can be cultured on a paste of cooked cereals in water instead of agar, e.g. *Panagrellus redivivus* (Winkler and Pramer 1961).

Many marine nematodes can be cultured xenically in agar: for example, nematodes from mangrove mud (Hopper *et al.* 1973), temperate brackish estuaries (Chitwood and Murphy 1964; Tietjen 1967), and shallow coastal water (Tietjen *et al.* 1970; Tietjen and Lee 1972, 1977*a*, *b*; Gerlach and Schrage 1971, 1972). *Rhabditis marina* and several genera of Monhysteridae (*Monhystera, Theristus, Diplolaimella* and *Diplolaimelloides*), can be cultured in sea-water nutrient agar, or sea-water/cereal cultures, together with a mixed bacterial flora. Algal-feeding nematodes, such as Chromadoridae are, in my experience, more difficult, the problem being to culture the appropriate algae. Lee, Tietjen, Stone, Muller, Rullman, and McEnery (1970) were able to culture a number of Chromadoridae from epiphytic communities of marine algae on bacteria. *Chromadora macrolaimoides* fed on bacteria and algae in culture (Tietjen and Lee 1973), but only two species of diatoms (*Nitzschia acicularis* and *Cylindrotheca closterium*) sustained culture indefinitely. Hopper and Meyers (1966) studied a variety of nematodes on fungal-infected cellulose in sea water.

Riemann and Schrage (1978) believe, from observations of cultures, that many marine nematodes from several families are mucus trap feeders, secreting mucus from the oesophageal or caudal glands and ingesting this with adherent bacteria, detritus and perhaps absorbed nutrients. Monhysteridae, Chromadoridae, and Oncholaimidae from the Weser Estuary were implicated in mucus feeding. Oncholaimidae are diverse in their feeding habits. *Oncholaimus oxyuris*, a brackish-water species, grew on a diet of bacteria and algae when small, but fed on nematodes (*Panagrellus redivivus*) when adult (Hiep *et al.* 1978). Lopez, Riemann, and Schrage (1979) believe dissolved organic matter from microbial activity may also be assimilated by *Adoncholaimus thalassophygas*. Perhaps the attraction of adult *Metoncholaimus* (from sea-grass *Thalassia*) to fungal mycelium, observed by Meyers and Hopper (1966), assures their larval progeny soluble nutrients. Chia and Warwick (1969) showed that *Pontonema vulgaris* assimilated (6-^{3}H) D-glucose from sea water, but the question arises as to whether such levels of glucose occur in the natural environment. *Pontonema vulgaris*, an intertidal species, readily fed on clotted blood in culture (Jennings and Colam 1970).

The identification of food and quantitative studies on nutrition in xenic cultures have been attempted by labelling food organism with ^{14}C or ^{32}P. The rate of ingestion of a number of marine bacteria and algae by *Rhabditis marina* (a widespread littoral nematode) was measured using

[32]P. The bacterium *Pseudomonas* sp. and the algae, *Nannochloris* and *Dunaliella*, were taken up in greatest numbers (Tietjen *et al.* 1970). Feeding by *Rhabditis marina*, *Monhystera denticulata*, *Chromadorina germanica*, and *C. macrolaimoides* on ^{32}P-labelled bacteria and algae (diatoms and chlorophytes) was compared (see Section 8.6) in a subsequent paper (Tietjen and Lee 1977*b*).

Some Tylenchida and Aphelenchida are easily cultured in agar media favouring fungal growth, such as Czarpek-Dox, potato dextrose, or cornmeal agar, together with the fungi on which they feed, puncturing the mycelium with their stylet. Relatively few genera of Tylenchida feed on fungi, which is commoner in the Aphelenchida. Some species may feed on either fungi or higher plant tissue. *Neotylenchus*, *Ditylenchus*, *Deladenus*, *Aphelenchus*, *Bursaphelenchus*, *Paraphelenchus*, and *Aphelenchoides* are common fungal-feeding genera (Hechler 1962*a*, *b*; Franklin and Hooper 1962; Faulkner and Darling 1961; Dropkin 1966; Younes 1969; Bedding 1968; Pillai and Taylor 1967). *Deladenus* species feed on fungi associated symbiotically with wood boring insects, and are facultative insect parasites (Bedding 1972). A Rhabditid, associated with bark beetles can be cultured on fungal conidia (Hunt and Poinar 1971).

Fungi differ in their suitability as food with species of *Pyrenochaeta*, *Alternaria*, *Rhizoctonia*, *Botrytis*, *Fusarium*, and *Agaricus* having proved successful in cultures. Nickle and McIntosh (1968) compared the reproduction of several different nematodes on different fungi and the effect the nematodes had on the fungi. Riffle (1968) found wide variation in the susceptibility and food value of many different mycorrhizal fungi. Faulkner and Darling (1961), using monoxenic cultures, found *Ditylenchus destructor* will feed and reproduce on 64 species (40 genera; 8 orders) out of 115 species of fungi tested, mostly common soil inhabitants, and will also live on higher plant tissues or plant callus cultures. Phycomycetes were not well represented with no Mucorales or Peronosporales in the 64. *Aphelenchus avenae* can feed on 43 different fungi (Townshend and Blackith 1975).

A marine species of *Aphelenchoides*, collected from submerged wood, feeds on numerous marine fungi, but as with terrestrial fungi and nematodes, these vary greatly in their capacity to support cultures of the nematode (Meyers, Feder, and Tsue 1963). *Zalerion xyelstrix*, *Halosphaeria mediosetigera*, and *Dendryphiella arenaria* supported good cultures.

4.3. Monoxenic cultures

Many of the bacteria-feeding nematodes referred to under the previous heading can be cultured monoxenically, if first freed from microbial contaminants, then transferred to pure cultures of a selected bacterial species in agar. Studies of bacteria as food show that not all bacteria are suitable (Sohlenius 1968). The most commonly used, for monoxenic culture, has been *Escherichia coli*. It has often been used to culture Rhabditidae and Cephalobidae: *Caenorhabditis briggsae*, *C. elegans*, *Rh. pellio*, *Pelodera strongyloides*, *Panagrellus redivivus*, *Acrobeloides buetschlii*, *Mesodiplogaster lheritieri* (Dougherty 1953; Yarwood and Hansen 1968; Cryan, Hansen, Martin, Sayre, and Yarwood 1963; Nicholas 1962; Brenner 1974; Sohlenius 1968; Nicholas, Grassia, and Viswanathan 1973; Anderson and Coleman 1981). *Bacillus cereus* has been used to culture a number of sand dune nematodes (Yeates 1970), while *Sarcina lutea*, *Staphylococcus aureus*, *Achromobacter aerogenes*, *Pseudomonas*, *Proteus*, *Arthrobacter*, and *Flavobacterium* have also been used for monoxenic cultures of Rhabditida (Nicholas 1962; Chin 1976; Sohlenius 1968; Anderson and Coleman 1981).

For studying the developmental biology and genetics of *Caenorhabditis*, many workers have followed the techniques of Brenner (1974). The nematodes are cultured on a 'lawn' of *E. coli* on a simple nutrient agar medium. A mutant *E. coli*, OP50, which requires uracil has been used because this requirement restricts the growth of the bacterium. Patel and McFadden (1976) used an isocitrate lyase-deficient mutant. Nematodes require sterols in their diet, and *E. coli* does not meet this requirement, but sufficient are present in nutrient agars containing animal or plant extracts, so that this is not a problem.

Chantanao and Jensen (1969) found that *Mesodiplogaster* (= *Pristionchus*) *lheritieri* could be cultured on a number of pathogenic bacteria from commercial crop plants, *Agrobacterium tumefasciens*, *Erwinia carotovora*, *E. amylovora*, and *Pseudomonas phaseolicola* and *Rhizobia* (Jatala, Jensen, and Russell 1974). Since this soil rhabditid is inefficient in digesting the bacteria ingested, passing some viable cells through its gut, it can disseminate these pathogens and symbionts (Jatala *et al.* 1974). Incidentally, this same rhabditid passes green algae (Leake and Jensen 1970), viable fungal spores (Jensen 1967; Jensen and Siemer 1969), bacteria phage (Jensen and Gilmour 1968), and mycoplasma (Jensen and Stevens 1969) through its gut. Its biology in culture has been studied by Grootaert (1976) and Anderson and Coleman (1981). Bacteria-feeding nematodes generally seem to pass some of the ingested bacteria through the gut unharmed, so that their dispersal and protection from chlorination become a public health concern (Chang, Berg, Clarke and Kabler 1960).

Establishing monoxenic cultures necessitates the initial freeing of the nematode from unknown contaminating micro-organisms (axenization). For rhabditids, which harbour viable micro-organisms in their alimentary canal, this can usually be accomplished by repeated washing in antibiotic mixtures, sometimes aided by bacteriostatic or bacteriocidal chemicals such as chlorine. Other workers have used toxic chemicals which kill the nematodes as well as the microbial contaminants, but not their eggs. Tylenchids and Aphelenchids are less difficult, because the stylet does not allow bacteria to enter the gut. For many purposes, monoxenic cultures do not offer any advantages over xenic cultures for fungal-feeding and predaceous nematodes, because their feeding habits are selective. Monoxenic cultures of fungal-feeding nematodes have been referred to in the previous section (Faulkner and Darling 1961; Dropkin 1966).

4.4. Axenic culture

Zimmerman (1921) may have been the first person to establish germ-free animals. He reported the axenic culture of the vinegar eelworm, *Turbatrix aceti* (= *Anguillula oxophila*) following treatment with H_2O_2, in a glucose, peptone, crude egg lecithin, NaCl medium. I was able to repeat his method of axenization (Nicholas 1956), but did not get sustained reproduction. Ells and Read (1952) similarly failed to establish sustained reproduction in their medium though they reported success when lecithin was replaced with an extract of liver. I failed with their medium too, but cultured *T. aceti* axenically for many generations (Nicholas 1959) using a culture medium based on that developed by Dougherty and Calhoun (1948) for *Caenorhabditis briggsae*. Metcalf (1903) had much earlier reported the culture of *Rhabditis brevispina* in sterile agar containing partly decomposed asparagus juices, but the evidence for the absence of microbial contaminants is not very convincing.

Success with axenic culture requires proof that contaminating microorganisms are not present and that the growth and reproduction of the organism can be sustained indefinitely. In my view, the first undoubtedly successful axenic cultures of nematodes were those of two facultative insect parasites *Neoaplectana chresima* and *G. glaseri* (Glaser 1940; Glaser, McCoy, and Girth 1942; Stoll 1959), using sterile unheated vertebrate tissues. These nematodes feed on bacteria in the putrefying cadavers of their insect hosts. It is questionable whether such associations are properly described as parasitic. Their work influenced Dougherty and Calhoun (1948) in establishing axenic cultures of *Caenorhabditis briggsae* (= *Rhabditis briggsae*). Successful methods for the axenic culture of *C. briggsae* were soon applied to related Rhabditida (see Table 4.1) but attempts to replace unstable tissue extracts with stable commercial products proved much more difficult (see reviews by Dougherty, Hansen, Nicholas, Mollett, and Yarwood (1959) and Nicholas *et al.* (1959) for early work). Table 4.1 also shows species more recently established in axenic culture, all from the Rhabditida or Aphelenchida, none from the Adenophorea. All the Rhabditida are bacteria-feeders, whilst the two

TABLE 4.1
Initial reports of successful axenic cultures of free-living nematodes

Species*	Repro-duction**	Origin	Reported by
Rhabditida			
Rhabditidae			
Caenorhabditis briggsae (= *Rhabditis briggsae*, = *R. elegans*)	H	Soil, California	Dougherty & Calhoun (1948)
C. elegans Bergerac	H	Soil, France	Nicholas *et al.* (1959)
C. elegans Bristol	H	Mushroom compost, England	Nicholas *et al.* (1959)
R. anomala (= *R. maupasi*)	H	Earthworm coelom, England	Nicholas and McEntegart (1957)
R. maupasi	B	Snails, Morocco	Brockelman and Jackson (1978)
Pelodera strongyloides	B	Bovine dermatitis	Yarwood and Hansen (1968)
Rhabditis marina	H	Marine, Long Island, NY	Tietjen and Lee (1975)
R. oxycerca (= *Pelodera* sp.)	B	Freshwater mud, ACT Australia	Marchant and Nicholas (1974)
Diplogasteridae			
Diplogasteroides	?	Freshwater, Illinois	Chaudhuri (1964)
Diplogaster nudicapitatus	?	Freshwater, Illinois	Chaudhuri (1964)
Cylindrocorpidae			
Cylindrocorpus longistoma	B	Sewage, Illinois	Chin (1976)
C. curzii	B		Chin (1976)
Cephalobidae			
Turbatrix aceti	B	Vinegar, England	Nicholas (1956)
Acrobeloides beutschlii	H	Peat, England	Nicholas (1962)
Panagrellus redivivus	H	Agricultural Exp. Station Florida	Cryan *et al.* (1963)
Chiloplacus lentus	?	India?	Roy (quoted from Vanfleteren (1978))
Aphelenchida			
Aphelenchoididae			
Aphelenchoides rutgersi (= *A. sacchari*)	P	Citrus roots, Florida	Myers (1967*a*)
Aphelenchidae			
Aphelenchus avenae	P or B	California	Hansen *et al.* (1970)

* Where the name used in the original paper has been changed, the original name is given in brackets.

** H means mostly protanderous hermaphrodites; B means bisexual; P means parthogenetic.

Aphelenchida feed on fungi not bacteria. The culture of these two aphelenchids in liquid culture media is very interesting because they feed through a stylet and the sensory stimuli associated with puncturing fungal hyphae and sucking out the contents has proved unnecessary for culture. One marine species, *Rhabditis marina*, is included.

Axenization of free-living nematodes using antibiotics and bacteriacidal chemicals is not difficult, with confirmation that microbial contamination is no longer present based on the absence of microbial growth when several standard microbiological culture media have been inoculated with media from the nematode cultures. Such methods do not entirely rule out microbial contamination, for example, by fastidious symbiotic micro-organisms, mycoplasmas, or viruses. Electron microscopy has not demonstrated such organisms in *Caenorhabditis* or the other important cultured species. Vanfleteren (1980) has examined stock cultures of *C. elegans* and *C. briggsae* for mycoplasmas, using selective culture techniques, without finding evidence for their presence.

Dougherty (1959) suggested the terms holidic, meridic, and oligidic to designate the stages he envisaged in the progressive refinement in culture techniques. A holidic medium is one composed of known chemicals, a meridic medium one with a chemically defined medium supplemented with some undefined material such as a tissue extract, and an oligidic medium is one that is chemically undefined. Progress in determining the nutritional requirements of nematodes has followed this pattern, so that the species referred to in Table 4.1 can be cultured in a meridic medium, with *Caenorhabditis* and *Turbatrix* also culturable in holidic media. For biochemical studies, however, what is usually required is enormous numbers of axenic nematodes in a cheap, easily prepared, oligidic medium.

C. briggsae, collected by Briggs from Californian soil in 1946 and put into axenic culture by Dougherty and Calhoun (1948), was used for nutritional studies, but as I mentioned in the 'Introduction', cultures of *C. briggsae* were later confused with those of *C. elegans*, and many papers on the nutrition and biochemistry of *C. briggsae* may in fact report work on *C. elegans*. Monoxenic cultures of *C. elegans* have been widely used for developmental biology and genetics (see Chapter 7). The differences between the two species have been clarified by Friedman, Platzer, and Eby (1977). *C. briggsae* (Dougherty and Nigon 1949; Dougherty 1953) differs from *C. elegans* (Maupas 1900; Dougherty 1953) in the arrangement of the bursal rays in the male. However, males are very infrequent in most populations of these predominantly hermaphrodite species. Friedman *et al.* (1977) examined cultures of *C. briggsae* from five different laboratories in different parts of the world, using controlled matings with a strain of *C. elegans* giving many males, and applying electrophoresis to their malic dehydrogenase isozymes and other proteins. The males of only one strain had the genuine *C. briggsae* bursal ray pattern; the other four resembled *C. elegans*. Incapacity to interbreed and electrophoretic mobility of proteins, including MDH-isozymes consistently confirmed the identity of *C. briggsae* and *C. elegans* as separate species. *C. briggsae* from Dr B. Zuckerman, University of Massachusetts, turned out to be authentic, the others, including mine, all derived from Dr E. C. Dougherty's laboratory in the first instance, were *C. elegans* (Friedman *et al.* 1977; Vanfleteren 1980).

Three different methods of assessing the nutritional value of culture media have been used. Dougherty and his co-workers (Dougherty and Hansen 1956; Dougherty *et al.* 1959) measured the generation time of individual nematodes. The eggs of *Caenorhabditis* sp. stick together in clumps in axenic culture media, and these were picked out, washed, and allowed to hatch overnight in 0.07 M potassium phosphate buffer (pH 7). The freshly hatched larvae were transferred singly to small silicone stoppered test tubes with 0.25 or 0.3 ml of medium. They were examined daily and the time for the completion of the first two generations (F1 and F2 times respectively) was estimated from the first appearance of progeny. The generation time is a good measure of the adequacy of the medium, being about 3–5 days in good media at 20 °C, though *C. briggsae* can lay eggs with 2½ days in monoxenic cultures with *E. coli*. Lower, Hansen, and Yarwood (1966) discussed statistical treatment of the data, recommending three larvae per tube to reduce variance, and plotting the regression of the reciprocal of generation time on log concentration of nutrient. For bisexual species 10 freshly hatched larvae can be used to inoculate each tube. In species in which eggs do not clump, the adult female can be washed and allowed to lay eggs in the buffer solution.

A different approach, used in several laboratories, estimates the total population produced in a given time in a larger volume of culture medium. Lu, Hugenberg, Briggs, and Stokstad (1978) assessed the utilization of fatty acids by *C. briggsae* from the population developing in 5 ml of medium after 28 days at 20 °C, starting from 500–900 nematodes. As time passes, the need for oxygen and accumulation of ammonia may limit growth cultures, while population structure changes with the accumulation of non-developing larvae, dead worms, and unhatched eggs, so that the conditions and length of culture are critical. Lu *et al.* (1978) cultured *C. briggsae* in larger test tubes, 18 × 150 mm, on an apparatus which rotated the tubes once a minute to facilitate gaseous exchange. Alternatively, one can measure populations after shorter intervals, while still growing rapidly. Thus Rothstein and Coppens (1978) used a similar method to Lu *et al.* (1978) but stopped cultures after nine days. In effect, it is the relative growth rate that is being compared, while cultural conditions are still favourable for rapid growth. Watson, Pinnock, Stokstad, and Hieb (1974) describe a nephelometer with which to monitor the growth of the population within culture tubes. A photocell measures the light scattered by the population. The population growth of *Caenorhabditis* and *Turbatrix* follows a classical sigmoid curve with population structure remaining constant during the logarithmic phase of rapid increase.

Each of these methods have proved sensitive measures of the nutritional value of culture media in experienced hands. Specific nutritional requirements will be discussed in Section 4.5 but I must turn now to the question of the need for protein in cultures, since this has been a major source of controversy. Meridic media consist of a chemically defined basal medium to which a partially purified protein-rich supplement must be added to get any of the nematodes to reproduce in axenic culture. The first chemically defined basal medium, GM, which was used for several rhabditids, was based on the dietary requirements of the cockroach (Dougherty and Hansen 1956; Dougherty *et al.* 1959). This was replaced by EM, in which the amino acids were based on the composition of *E. coli* (Sayre, Hansen, and Yarwood 1963). EM was further modified (Buecher, Hansen, and Yarwood 1966) becoming the basis for a commercially marketed medium, CbMM, or *C. briggsae* maintenance medium, prepared and sold by Grand Island Biological Company (GIBCo), in the USA and UK (Table 4.2).

CbMM has been widely used for axenic cultures of nematodes. It omits a number of nutrients, the need for which has become apparent since it was formulated, and contains others which are clearly redundant. For example, it contains multiple forms of the same vitamins, while others such as folic acid and cyanocobalamin are at many times their proven requirements. A more rational medium, EMS, has been produced by Rothstein (1974). Nonetheless, CbMM, because of its wide usage and ready availability, has proved very useful. Generally, a sterol, e.g. β-sitosterol or cholesterol, and a source of haem, e.g. myoglobin or haemin chloride, are added to give a meridic basal medium of wide applicability for nematode culture. It has formed the basis for meridic cultures of *C. briggsae*, *C. elegans*, *Panagrellus redivivus*, *Turbatrix aceti*, *Aphelenchoides rutgersi*, *Aphelenchus avenae*, *Neoaplectana carpocapsae*, and *N. glaseri*.

It was found that the supplements required in meridic media could be prepared from a variety of vertebrate tissues as well as bacteria (Nicholas *et al.* 1959). Liver protein fractions from several mammals were mostly used, though unfractionated chick embryo extract was also very effective. It is now clear that these extracts provided at least three different requirements. Hieb and Rothstein (1968) found that *E. coli*, in a simple salt solution, provided an inadequate medium for the monoxenic culture of *C. briggsae*, but were adequate with agar or added sterols. Sterols required by nematodes will be discussed more fully later (Section 4.5). Autoclaved *E. coli*, unlike unheated *E. coli*, in a buffered salt solution are inadequate, even with the addition of sterols. However, this axenic medium becomes adequate with the addition of a lamb's liver extract (Hieb, Stokstad, and Rothstein 1970). Further study showed that the active fraction contained a haemprotein, probably myoglobin rather than cytochrome or haemoglobin. Myoglobin, cytochrome-c, or haemin chloride all gave good cultures when combined with autoclaved *E. coli* and sterols in buffered salt solution. Thus two of the necessary constituents of tissue extracts were identified as sterol and haem, but these, when added to a chemically defined basal medium, still remained insufficient for sustained growth and reproduction of *Caenorhabditis*.

Critical reviews by Vanfleteren (1978, 1980) discuss, with the benefit of hindsight, the many attempts to resolve the nature of the remaining undefined requirements met by liver, embryo, yeast, and bacterial extracts. Several hypotheses have been advanced to account for the need for protein-rich tissue supplements, generally heat-labile, though apparently heat-stable in *E. coli*. A partially purified liver fraction, LPF-C, showed enhanced potency when frozen and thawed in the presence of other media constituents, notably nucleotides, i.e. freeze-activation (Hansen, Sayre, and Yarwood 1961; Hansen, Buecher, and Yarwood 1964). It was suggested that the active constitutent, GF, was a protein, which could exhibit different degrees of aggregation and which was present in either active or inactive form in various tissues (Sayre, Lee, Sandman, and Perez-Mendez 1967).

An alternative hypothesis is that *Caenorhabditis* requires peptides, as distinct from amino acids. The addition of hydrolysed casein, hydrolysed soy protein, and several specific di- and tri-peptides stimulates the growth of *C. briggsae*, as does unhydrolysed casein, when added to meridic media based on CbMM (Pinnock, Shane, and Stokstad 1975). However, it has been possible to culture *C. briggsae* in CbMM + glycogen + haemin (Hansen, Perez-Mendez, and Buecher 1971).

Vanfleteren (1975*a*, *b*, 1976*a*, 1978, 1980) argues that the heat-labile supplements give a fine-precipitate in the medium which serves as a carrier for haem (see also Buecher, Perez-Mendez, and Hansen 1969). Thus a wide variety of different supplements can fulfil the requirement (in the presence of haem and sterol and CbMM). It accounts for the growth promoting activity of ribonucleoproteins and various non-haem proteins in the presence of haemin, under appropriate conditions (Vanfleteren 1975*a*); and for the effectiveness of the haem proteins, haemoglobin, myoglobin, cytochrome, peroxidase, and catalase, in the absence of haemin, when properly precipitated (Vanfleteren 1975*b*); and for certain artificial haemin-protein complexes (Vanfleteren 1976*a*). However, Vanfleteren's hypothesis has not been accepted by all workers in the field.

Rothstein and Coppens (1978) obtained very good cultures of *C.*

briggsae, *C. elegans*, and *Panagrellus silusiae* with a basal medium (EMS, Rothstein 1974), soy peptone, and haemin chloride. Myoglobin was also used in place of haemin. They partially purified and identified the active fraction from soy pepone as a polypeptide. Lu *et al*. (1978) re-examined *C. briggsae*'s need for lipids, using CbMM + cytochrome-c + sterols + Tween 80 (polyoxyethylene sorbitan mono-oleate) (Sigma, St. Louis, USA, an emulsifying agent for sterols). They found Na oleate or K acetate stimulated growth, and reported successful culture through a number of subcultures with a chemically defined medium, i.e. CbMM + 50 μg/ml cytochrome-c + 50 μg/ml β-sitosterol + 0.5 or 1 mg/ml Na oleate (8 serial sub-cultures) or with 0.5 mg/ml K acetate in place of Na oleate (4 serial cultures). Yields of 80 000–120 000 nematodes per ml in four weeks in the former, and 60 000–70 000 per ml in the latter, was reported.

The controversy surrounding the need for specific proteins or peptides, and for precipitated proteins as a carrier for haemin remains unsettled. It is possible that continuous *in vitro* culture for 30 years has led to the selection in some laboratories of strains adapted to soluble nutrients. It is clear that haem proteins are more readily utilized than free haemin, but that heat denaturation makes haem proteins unavailable. We need to know more about the mechanisms of assimilation in the nematode intestine and Vanfleteren (1980) has reported some preliminary observations by electron microscopy on the assimilation of precipitated ferritin (an iron-containing protein). Possibly proteins are taken up by the intestinal cells of *Caenorhabditis* by pinocytosis and digested intracellularly. Conditions stimulating pinocytosis would be essential for assimilation.

Most of the experiments referred to above have been made on *Caenorhabditis*, reported as *C. briggsae* though in many cases *C. elegans*, as already explained. Experiments with other species have progressed to very varying degrees, but tend to suggest that the Rhabditidae are rather similar in cultural requirements. *R. anomala*, *R. maupasi*, and *P. strongyloides* have been cultured in ologidic media with liver extracts or chick embryo extracts (Rothstein and Cook 1966; Brockelman and Jackson 1978; Yarwood and Hansen 1968, respectively). Of the Cephalobidae, *Acrobeloides beutschlii* can be cultured in autoclaved liver extract plus chick embryo or heated liver extract (HLE) (Nicholas 1962). The original paper on *Chiloplacus lentus* (Roy quoted from Vanfleteren 1978) was unavailable to me (see Table 4.1). *Panagrellus redivivus* (Cryan *et al*. 1963) and *Turbatrix aceti* have been more widely studied; *T. aceti* especially for experiments on metabolism. *T. aceti* and *P. redivivus* produce good cultures in the oligidic media developed for *C. briggsae* (Rothstein and Cook 1966). *T. aceti* gives good cultures in a defined basal medium (EMS) + 50 μg/ml cholesterol + 4 per cent acetic acid + 500 μg/ml of myoglobin or haemoglobin (Rothstein 1974). Acetic acid stimulates cultures of *T*.

TABLE 4.2

Caenorhabditis briggsae *maintenance medium (Hansen 1966)*

Component	mg/l	Component	mg/l
$CaCl_2\,2H_2O$	220.5	L-serine	788.0
$CuCl_2\,2H_2O$	6.5	L-tyrosine	272.0
$MnCl_2\,4H_2O$	22.2	L-phenylalanine	180.0
$ZnCl_2$	10.2	Glutathione, reduced	204.0
KH_2PO_4	1225.5	N-acetylglucosamine	15.0
Potassium Citrate H_2O	486.0	Cyanocobalamine	3.75
$Fe(NH_4)_2(SO_4)_2\,6H_2O$	58.8	Folinate (Ca)	3.75
Magnesium Citrate $5H_2O$		Niacinamide	7.5
(Dibasic)	915.0	Pantetheine	3.75
KOH	*	Pantothenate (Ca)	7.5
D-glucose	1315.0	Pyridoxal phosphate	3.75
L-arginine	975.0	Pyridoxamine 2HCl	3.75
L-histidine	283.0	Pyridoxine HCl	7.5
L-lysine–HCl	1283.0	Riboflavin-5'-PO_4(Na) $2H_2O$	7.5
L-tryptophan	184.0	Thiamine HCl	7.5
L-phenylalanine	623.0	p-aminobenzoic acid	7.5
L-methionine	389.0	Biotin	3.75
L-threonine	717.0	Niacin	7.5
L-leucine	1439.0	Pterolyglutamic acid	7.5
L-isoleucine	861.0	DL-thioctic acid	3.75
L-valine	1020.0	Choline dihydrogen Citrate	88.5
L-alanine	1395.0	Myo-inositol	64.5
L-aspartic acid	1620.0	Adenosine-3'-(2')-phosphoric acid	
L-cysteine HCl H_2O	28.0	H_2O	365.0
L-glutamate (Na) H_2O	550.0	Cytidine-3'-phosphoric acid	323.0
L-glutamine	1463.0	Guanosine-3'-(2')-$PO_4Na_2\,H_2O$	363.0
Glycine	722.0	Uridine-3'-(2')-phosphoric acid	324.0
L-proline	653.0	Thymine	126.0

* As needed for adjustment to pH 5.9 ± 0.1.

aceti. Mention should also be made here of the insect parasites. *Neoaplectana glaseri* and *N. carpocapsae*, because their life-cycles resemble those of *R. anomala* and *R. maupasi*, which are facultative invaders of earthworms and snails. *N. carpocapsae* has been cultured in CbMM + human gamma-globulin + haemin, like *Caenorhabditis* and *P. redivivus* (Buecher, Hansen, and Yarwood 1970).

Rhabditis marina is of special interest because it is a marine nematode, though tolerant of a very wide range of osmotic pressures. Tietjen and Lee (1975) established axenic cultures for more than 75 generations in a complex defined medium, different from CbMM, supplemented with filtered (not autoclaved) sheep blood or haemoglobin. They reported that their cultures were accidentally destroyed before nutritional requirements could be analysed, though they had found lipid extracts stimulatory.

The stylet-feeding aphelenchids, *Aphelenchoides rutgersi* and *Aphelenchus avenae*, require more concentrated and different proteins to *Caenorhabditis*. *A. avenae* can be cultured in CbMM + 10 per cent human serum + 25 per cent chick embryo extract (Hansen, Buecher, and Evans 1970). *A. rutgersi* cultures well in the same medium (Myers, Buecher, and Hansen 1971), or in CbMM + 10 per cent chick extract plus soy peptone and yeast extract, or in very concentrated liver extracts (50 per cent HLE) (Myers 1967*b*). Limited growth and reproduction occurred in a holidic medium (Myers 1968).

A cheap medium giving good cultures of *C. briggsae, C. elegans, T. aceti,* and *A. buetschlii* consists of an autoclaved liver extract, ALE, supplemented with HLE or unfiltered chick embryo extract (Nicholas *et al.* 1959). A widely used cheap medium, in which several species will grow, is 4 per cent soy peptone, 3 per cent yeast extract, 10 per cent HLE; the first two are available commercially. HLE is prepared by filtering the supernatant from homogenized lamb's liver after heating for six minutes at 53 °C (Sayre *et al.* 1963). It is inactivated by autoclaving and is a necessary source of haem. Rothstein (1974) has evaluated media for *C. briggsae* and *T. aceti* and substituted haemoglobin, which is cheaper, for HLE. *T. aceti* requires acetic acid, 4 per cent or 0.2 ml glacial acetic for 5 ml medium. For the culture of very large numbers of *C. elegans* at minimum cost,

TABLE 4.3
Probable nutritional requirements of Caenorhabditis briggsae

Required	Stimulate growth
Amino acids	*Carbohydrates*
L-arginine	glucose
L-histidine	trehalose
L-lysine	
L-tryptophan	*Peptides*
L-phenylalanine	L-leucyl-L-phenylalanine
L-leucine	unidentified oligopeptide
L-isoleucine	
L-threonine	*Lipids*
L-methionine *or* homocysteine	Oleic acid *or* acetate
Sterols	*Others*
Cholesterol *or* one of many	acetate *or* ethanol *or* oleic acid
alternatives	
	Inorganic
Fe-porphyrin	Magnesium
haemin *or* haemprotein	Sodium
	Potassium
Vitamins	Manganese
thiamine	Calcium
riboflavin	Copper
niacinamide	Chloride
pantotheine	
pyridoxine	
folic acid	
biotin	
cyanocobalamine	

Vanfleteren (1976*b*) has developed a medium with 3 per cent soy peptone, 3 per cent yeast extract (low in NaCl), 1 per cent dextrose, 1 per casein hydrolysate and 500 μg/ml haemoglobin. Rothstein and Hansen and their co-workers use United States suppliers; Vanfleteren uses European suppliers.

Cultures produce high populations when shallow, because of the need for oxygen, and gaseous exchange also reduces the accumulation of ammonia in the medium. This can be achieved with shallow media in large Erlenmeyer flasks, fitted with cotton stoppers or loose-fitting metal caps, or more numerous 18 × 150 mm screw-capped tubes, each with 5 ml of medium, on a tissue rotator (at 1 r.p.m.) (Lu *et al.* 1978). Hansen and Cryan (1966*a*) described a method in which a thin film of medium is spread over glass wool and drained at intervals, giving a continuous supply of nematodes. Populations can be increased by bubbling air, sterilized by filtration through millipore membranes, through the medium, but an antifoaming agent must be included. Buecher and Hansen (1971) reported increased yields of *C. briggsae*, *P. redivivus*, *T. aceti*, and *N. carpocapsae* in this way. With *C. elegans* 1 g of lyophilized nematodes was obtained from 500 ml of medium; over 100 000 per ml. Vanfleteren (1976*b*) obtained 10^9 *C. elegans*, or 40 g wet weight, in six weeks from 50 nematodes, using a spinner flask, aeration, and the cheap medium described above (plus an antifoaming agent).

4.5. Nutritional requirements

Many of the minimal nutritional requirements of *C. briggsae* have been established by observing the effects of specific omissions from axenic culture media (see Table 4.3). However, specific requirements for one nutrient may depend on the availability of other specific nutrients, or the overall balance of nutrients. Monoxenic cultures have also been used effectively to show requirements, as with sterols. Radioisotopes have been used to demonstrate synthetic capabilities, for example, with amino acids (Nicholas, Dougherty, Hansen, Holm-Hansen, and Moses 1960; Rothstein and Tomlinson 1961, 1962; Balasubramanian and Myers 1971), but these have not proved a reliable guide to nutritional requirements. Thus despite some of the amino acids becoming labelled in *C. briggsae* from ^{14}C-labelled precursors (Rothstein and Tomlinson 1961, 1962) they are still required in the diet. A similar disparity was reported by Balasubramanian and Myers (1971).

Vanfleteren (1973) found that by single omissions from media, *C. briggsae* required arginine, histidine, lysine, tryptophan, phenylalanine, methionine, threonine, leucine, isoleucine, and valine, whilst alanine, asparagine, cysteine, glutamate, glutamine, glycine, proline, serine, and tyrosine were non-essential. Homocysteine can be converted to methionine (Lu, Hieb, and Stokstad 1976) and some simple peptides stimulate growth, even in the presence of high levels of amino acids (Pinnock, Shane, and Stokstad 1975; Rothstein and Coppens 1978). *N. glaseri* requires the same essential amino acids as *C. briggsae* (Jackson 1973), while tyrosine is beneficial and aspartic acid can be toxic. With *A. rutgersi* the situation is less clear (Balasubramanian and Myers 1971), however, those probably required are cysteine (cystine), methionine, histidine, phenylalanine + tyrosine, tryptophan, isoleucine, and leucine.

Carbohydrate requirements have been less studied, but glucose or trehalose sustained reproductive levels when included in the basal medium for *C. briggsae*, *C. elegans*, and *R. anomala* (though *P. redivivus* was unaffected) (Hansen and Buecher 1970). With *A. rutgersi*, glucose, raffinose, ribose, mannose, and trehalose were beneficial in a holidic medium (Petriello and Myers 1971).

A requirement for sterols was first shown in *N. carpocapsae* by Dutky, Robbins, and Thompson (1967), and then, as described in the previous section, in *C. briggsae* by Hieb and Rothstein (1968). Cholesterol, cholestane, 7-dehydrocholesterol, ergosterol, β-sitosterol, and stigmasterol can satisfy this requirement. Agar usually contains sufficient to meet this need, which in *C. briggsae* may not exceed 1.3 μg per ml. In *C. briggsae*, *C. elegans*, and *T. aceti* 0.1–2 μg per ml of β-sitosterol is sufficient (Lu, Newton, and Stokstad 1977). It is possible to show a need of sterols in monoxenic cultures, though this may not show without subculturing. Cole and Dutky (1969) demonstrated such a requirement in *T. aceti* and *P. redivivus*, on a diet of *Bacillus subtilis* endospores in glucose–peptone agar. For *T. aceti* the following satisfied the requirement: cholesterol, 7-dehydrocholesterol, Δ^7-cholestenol, cholestanol, β-sitosterol, fucosterol, 24-methylene-cholesterol, 25-norcholesterol, cholest-4-ene-3-one, campesterol, stigmasterol, and desmosterol, but not coprostanol or coprost-7-ene-ol. Between 0.1 mg per ml and 10 μg per ml was required. *P. redivivus* utilized cholesterol.

The demonstration that haem proteins or haemin are required by *C. briggsae*, *C. elegans*, and *T. aceti* has already been discussed (Section 4.4). The requirement is for Fe-protoporphyrin, either bound to protein or unbound as in haemin. Myoglobin, haemoglobin, cytochrome-c (not perhaps for *T. aceti*), and peroxidase catalase can meet this need when properly prepared (Vanfleteren 1976*a*).

The systematic omission of B-vitamins (in the presence of tissue supplements) indicates a requirement in *C. briggsae* for thiamin, riboflavin, folic acid (= pteroylglutamic acid), pyridoxine, niacinamide, and panthothenic acid (Nicholas, Hansen, and Dougherty 1962). A folic-acid antagonist, aminopterin, inhibits reproduction (Dougherty and

Hansen 1957) and creates a need for thymine (Vanfleteren and Avau 1977). In the absence of folic acid, formimino-L-glutamic acid accumulates in the tissues (not the medium) and histidine degradation is impaired (Lu, Hieb, and Stokstad 1974). The optimal levels for folic acid per ml, and for vitamin B_{12} (cyanocobalamin), of 6 ng-ml, were determined in *E. coli* supplemented media by Lu *et al.* (1976). Vitamin B_{12} and folic acid are required for the conversion of homocysteine to methionine, while choline was not needed, so that methylation of homocysteine is probably the pathway involved. Avidin, a protein that binds biotin, inhibits the growth of *C. briggsae* when added to culture media, but this is reversible (Nicholas and Jantunen 1963). A biotin requirement is probable, though it is possible that avidin is assimilated unaltered to bind biotin synthesized by *C. briggsae*.

Despite its capacity to synthesize oleic acid and polyunsaturated fatty acids from [^{14}C-2] acetate (Rothstein 1980), fatty acids stimulate the growth of *C. briggsae* and may be required (Lu *et al.* 1978). In a holidic medium, Na oleate and K acetate were required for continued subculture. Several compounds stimulated growth in place of oleate or acetate, but were not tested in serial subculture, i.e. Na stearate, polyoxyethylene sorbitan mono-oleate (Tween 80), or trioleate (Tween 85), n-propanol, and ethanol. Na linoleate was less active than oleate, suggesting monounsaturated or saturated fatty acids are better than polyunsaturated acids. Tween 80 (Sigma, St. Louis, USA) is added to most basal media to dissolve sterols and some vitamins prior to their addition. Stimulation of the growth of *C. briggsae* and *T. aceti* by acetate reinforces the importance of the glyoxylate cycle in nematodes (Section 5.6).

A. rutgersi requires haem or haemin and purines for good growth in a medium containing a depleted chick embryo extract (Thirungnam 1976). Guanosine or guanylic acid were needed for egg production, but adenine could not be tested because ATP was part of the basal medium. Pyrimidines appear to be more readily synthesized than purines, from experiments with [^{3}H-5] orotic acid (for pyrimidines) and [^{14}C-U] glycine (for purines) (Thirungnam and Myers 1974). Omitting purines and pyrimidines from CbMM reduces the growth of *C. elegans* in culture, which can be restored by adding adenine, adenosine, or AMP (Platzer and Eby 1980). Perhaps hypoxanthine phosphoribosyltransferase is lacking.

It has been possible to investigate the cations required by *C. elegans*, by the systematic omission and replacement of chloride salts, from a holidic medium (Lu 1980). Maximal populations were obtained with the following concentrations in µg per ml: Mg, 73; Na, 300; K, 530; Mn, 6.3; Ca, 1500; and Cu, 7.2. Zinc deficiency was not demonstrable and iron was present as haem or cytochrome. Myers (1971) investigated the physicochemical requirements for culturing *A. rutgersi*. Osmotic pressure may be significant, but viscosity did not appear to be, which is interesting because *A. rutgersi* ingests food through a narrow stylet canal.

5. Biochemistry

5.1. Introduction

It is the respiratory metabolism of parasitic nematodes from the alimentary canal of vertebrates which has attracted the most attention from biochemists. Like other alimentary-canal helminth parasites, their respiratory metabolism differs from the classical metabolic pathways of vertebrates, possessing different pathways for anaerobic respiration. The differences can be interpreted as adaptations to a largely anaerobic environment, but these helminths do utilize oxygen when it is available, and in some stages of the life-cycle, passed outside the host, must respire aerobically. Alternatively, one can interpret these helminth pathways as having evolved in free-living ancestral helminths adapted to widespread anaerobic environments. Nematodes dominate the marine meiofauna and are common in anaerobic environments of all kinds, for example, the 'sulphide' zone of marine and limnic sediments (Ott and Schiemer 1973). However, almost no work has been done on the biochemistry of nematodes from such anaerobic habitats, and such work as has been done on other free-living terrestrial nematodes does not point to the alternative anaerobic metabolism referred to above.

Biochemical studies have concentrated overwhelmingly on *Ascaris suum*, which is both parasitic and much larger than most other nematodes, giving a biased view of nematode biochemistry. However, recent dramatic advances in the genetics, neurobiology, and developmental biology of *Caenorhabditis* have interested a much wider spectrum of biologists in nematode biochemistry. There has been important work on the muscle proteins, neurotransmitters, and biochemical genetics. Nematodes have also proved useful for studies on the biochemistry of cryptobiosis, discussed in Section 3.8.

The importance of nematodes in plant pathology has stimulated work on the biochemistry of plant and fungal-feeding Tylenchida and Aphelenchida. Their digestive enzymes and lipid and nitrogen metabolism have been most fully investigated. The development of methods for axenic and monoxenic culture has drawn attention to the nutritional peculiarities of nematodes, leading in turn to studies of their capacities to synthesize or convert dietary fatty acids, sterols, and amino acids.

5.2. Lipids

Total lipids usually comprise about 25 per cent of the dry weight (Table 5.1) and lipids are a major food reserve. *Aphelenchus avenae* uses lipid

TABLE 5.1
The lipid content of nematodes

Species	Total lipids per cent dry weight	Food or medium	References
Turbatrix aceti	20.2–28.7	Axenic culture	Krusberg (1972)
,, ,,	29	Vinegar	Barrett *et al.* (1971)
,, ,,	19.8	Unstated	Chitwood and Krusberg (1980)
,, ,,	21.9	Vinegar	Womersley, Thomson, and Smith (1982)
Panagrellus redivivus	23	Oatmeal	Barrett *et al.* (1971)
,, ,,	24	Oatmeal	Sivapalan and Jenkins (1966)
,, ,,	23.14	Flour	Womersley *et al.* (1982)
,, ,,	15	Axenic culture	Willet and Downey (1973)
Caenorhabditis sp	33	Bacteria	Cooper and Van Gundy (1970)
Aphelenchus avenae	36	Fungus	Cooper and Van Gundy (1970)
,, ,,	36.5	,,	Womersley *et al.* (1981)
Ditylenchus triformis	20.8–24.9	,,	Krusberg (1967)
,, *dipsaci*	36.9–28.4	Alfalfa callus	Krusberg (1967)
,, ,, (L$_4$)	38.4	Narcissus bulb (dry)	Womersley *et al.* (1981)
Pratylenchus penetrans	25.3–28.4	Alfalfa callus	Krusberg (1967)
Tylenchorhynchus claytoni	34.3–39.7	Alfalfa callus	Krusberg (1967)
Aphelenchoides ritzemabosi	10.1–11.1	Alfalfa callus	Krusberg (1967)
Globodera solanacearum	29.4	Plant roots	Orcutt, Fox, and Jake (1978)
Meloidogyne javanica (♀)	40.5	Tomato roots	Chitwood and Krusberg (1981*a*)
Anguina tritici (L$_2$)	40.6	Wheat (dry)	Womersley *et al.* (1982)

reserves preferentially when starved under aerobic conditions, glycogen under anaerobic conditions, replacing the lost glycogen at the expense of lipids when returned to aerobic conditions (see Table 5.2). *Caenorhabditis* sp does not spare glycogen when starved under aerobic conditions and is more susceptible to anaerobiosis (Cooper and Van Gundy 1970). Glycogen is largely replaced by trehalose in anhydrobiotic nematodes (see Section 3.8) but lipids remain important reserves (Womersley, Thomson, and Smith 1982).

Interest has centred on the fatty acid composition of fats and phospholipids, and the variety of sterols present in nematode tissues. The number of carbon atoms and the number and position of double bonds (i.e. degree of unsaturation) are of interest, when compared with those in the diet, because this reflects the nematode's capacity to synthesize or modify

TABLE 5.2

The utilization of glycogen and lipid reserves under various conditions

Species	Conditions	Glycogen as dry weight (%)	Lipid as dry weight (%)	Reference
Panagrellus redivivus				
Unstarved	Aerobic	6.5	23	Barrett, Ward, and Fairbairn (1971)
Starved 10 days	Aerobic	3.7	18	Barrett *et al.* (1971)
Unstarved	Aerobic	25.23	23.14	Womersley, Thomson, and Smith (1982)
Turbatrix aceti				
Unstarved	Aerobic	19	29	Barrett *et al.* (1971)
Starved 10 days	Aerobic	11	25	Barrett *et al.* (1971)
Unstarved	Aerobic	22.73	21.9	Womersley *et al.* (1982)
Caenorhabditis sp.				
Unstarved	Aerobic	3.3–3.4	34–6	Cooper and Van Gundy (1970)
Starved 10 days	Aerobic	trace	11	Cooper and Van Gundy (1970)
Starved 10 days	Anaerobic	Did not survive		Cooper and Van Gundy (1970)
Aphelenchus avenae				
Unstarved	Aerobic	7.8–8.2	32–4	Cooper and Van Gundy (1970)
Starved 10 days	Aerobic	8.0	10	Cooper and Van Gundy (1970)
Starved 10 days	Anaerobic	1.8	33	Cooper and Van Gundy (1970)
Unstarved	Stored frozen	2.13	36.5	Womersley *et al.* (1982)
Anhydrobiotic	dry pellet	1.24	14.0	Womersley *et al.* (1982)

assimilated lipids. Nematodes require sterols in their diet, but the question remains as to the capacity of nematodes to modify those assimilated.

Fatty acid composition has been investigated in *Caenorhabditis* sp, *Panagrellus redivivus*, *Turbatrix aceti*, and *Aphelenchus avenae* (Cooper and Van Gundy 1971; Barrett, Ward, and Fairbairn 1971; Womersley, Thomson, and Smith 1982, respectively), as well as in several plant parasites. Oleic acid, C:18:1 (18 carbon atoms and one double or unsaturated bond), predominates with lineoleic, C:18:2 (two double bonds), the next most important. A number of other even carbon numbered fatty acids are present at lower concentrations with differing numbers of unsaturated double bonds. About 80 per cent of the total lipid is generally fatty acid and about 90 per cent of the fatty acids are present as triglycerides, with smaller proportions as di- and monoglycerides and as free fatty acid. In *T. aceti*, Krusberg (1972) identified 31 different fatty acids, 65 per cent unsaturated (29–36 monounsaturated; 26–39 polyunsaturated). The shorter acids, C-10 to C-17, were almost completely saturated, with traces of acids with odd-numbered C atoms. Oleic, *cis*-vaccenic, and linoleic acids were important constituents.

In five Tylenchida: *Ditylenchus triformis, D. dipsaci, Pratylenchus penetrans, Tylenchorhynchus claytoni,* and *Aphelenchoides ritzemabosi,* Krusberg (1967) found a great diversity of fatty acids and a greater variety of unsaturated fatty acids than in the plant callus culture or fungal mycelium on which they had been feeding. The host tissue could have supplied much of the saturated fatty acids, but most of the unsaturated fatty acids were believed to have been produced by the nematodes. Fatty acids for C-12 to C-22 were found, C-18:1 predominating in all five nematodes, but with the double bond between C7 and C8 (vaccenic), not C9 and C10 (oleic) as in the host. Polyunsaturated acids were present, C-18 with one to three, and C-20 with none to five double bonds, as well perhaps, as some branched chain fatty acids.

Globodera solonacearum, a plant root parasite, differs from the foregoing species in that C-20 fatty acids predominate over C-18 acids; C-20:1 and C-20:4 are the most important in neutral, free, and phospholipid fractions, with C-18:0 and C-18:1 of second importance. The parasitic females of *G. solonacearum* (Orcutt, Fox, and Jake 1978) and anhydrobiotic *Anguina tritici* and *A. avenae* (Womersley *et al.* 1982) gave the following lipid composition

Species	Triglyceride	Free fatty acids	Phospholipids	Paraffin hydrocarbon
Globodera solonacearum				
% total lipid	47.27	6.46	4.86	0.48
% dry weight	13.28	1.81	1.36	0.13
% unsaturated	77.95	77.75	74.33	—
Aphelenchus avenae				
% total lipid	79.0	12.8	8.2	—
Anguina tritici				
% total lipid	85.3	trace	14.6	—

The differences between the fatty acid found in nematodes and in their food strongly suggest that the nematodes can synthesize or modify dietary fatty acids. Rothstein and Gotz (1968) showed that *T. aceti* can synthesize fatty acids from [2-^{14}C] acetate in axenic culture. The pattern of labelling left no doubt that fatty acids were synthesized *de novo*. Polyunsaturated fatty acids were formed from saturated fatty acids, and the capacity of *T. aceti* to introduce a second, or further unsaturated double bonds, into monounsaturated fatty acids, distinguishes nematodes from higher animals. *C. briggsae* and *P. redivivus* can similarly synthesize fatty acids from

[2-^{14}C] acetate, with acids from C-14 to C-20 becoming labelled (in CH$_3$ as well as COOH terminal carbons). The metabolism of ^{14}C-labelled fatty acids showed that like *T. aceti*, these nematodes can unsaturate C-18:0 and C-18:1 to C-18:2. *P. redivivus* can probably lengthen dietary fatty acids as well as *de novo* synthesis. In the light of Rothstein's work, it is interesting to recall that acetate or monounsaturated oleic acid stimulates growth and may be a requirement for *C. briggsae*, while polyunsaturated acids are less efficient (Lu *et al.* 1978; see Section 4.5).

The phospholipids comprise an important fraction of the total lipids, and include a diverse mixture of both unsaturated and saturated fatty acids as diacyl phosphoglycerides. Phospholipids were reported to account for 32.8 per cent of total lipids in *P. redivivus* (Sivapalan and Jenkins 1966), but are usually less. Phosphatidyl ethanolamine and phosphatidyl choline are the major phospholipids present in *P. redivivus*, *T. aceti*, *A. avenae*, *A. tritici* and *M. javanica*, with small amounts of phosphatidyl inositol, phosphatidyl serine, and phosphatidic acid (Sivapalan and Jenkins 1966; Chitwood and Krusberg 1980, 1981*a, b*; Womersley *et al.* 1981). Chitwood and Krusberg (1980) found small percentages of phosphatidyl serine and sphingomyelin in *T. aceti*; and all these plus lysophosphatidyl choline in *M. javanica* (Chitwood and Krusberg 1981*a*).

The predominant fatty acid of phospholipids in all three species is oleic 18:1, with significant amounts of stearic 18:0. Generally, phospholipids contained more polyunsaturated fatty acids than neutral lipids, amongst the different fatty acids present in both fractions. Two analogues of diacylphosphaglycerides, namely 1-alkyl-2-acyl and 1-alkenyl-2-acyl (plasmalogen) ethanolamine, and choline phosphoglycerides have been found in *T. aceti* and *M. javanica*, the major fatty acid being 18:1 in both these analogues, accounting for nearly half the ethanolamine phosphoglyceride in *M. javanica*. The analogues are not common in plants, and were not present in the host plant of *M. javanica*, though known from other animals.

Total lipids comprised the following fractions

	Neutral lipid	Glycolipid	Phospholipid
T. aceti	44.1	4.5	51.4
M. javanica	84.2	2.5	13.3

Nutritional studies have shown that some nematodes require sterols in their diet (see Section 4.5), which lends interest to analyses of the sterols found in nematodes. In *Ditylenchus triformis and D. dipsaci*, cholesterol and lathosterol are the principal sterols (Cole and Krusberg 1967), with traces of the fungal and plant callus phytosterols found in their foods, respectively, also present. In *D. dipsaci* the figures were rather similar. In *T. aceti* cholesterol and 7-dehydrocholesterol are the major sterols with lathosterol a minor constituent (46, 44, and 10 per cent respectively; Cole and Krusberg 1968). Sterol content could be increased from 0.02–0.03 per cent of dry weight to 0.15 per cent by increasing sterols in culture medium. Sterols, seven of which were detected, were low in *G. solanacearum*, about 0.01 per cent of dry weight (Orcutt *et al.* 1978).

T. aceti, C. briggsae, and *P. redivivus* do not incorporate [2-^{14}C] acetate, or [2-^{14}C] DL-mevalonate into sterols when these are supplied in axenic culture (Rothstein 1968; Cole and Krusberg 1968). These are precursors of the initial steps in sterol synthesis. *P. redivivus* converts [4-^{3}H] squalene 2, 3 oxide from the culture medium to lathosterol, implying that 2, 3 oxidosqualene cyclase is present, so that one of the terminal steps in synthesis is accomplished, though the conversion of lathosterol to cholesterol does not occur (Willett and Downey 1973, 1974). It seems that the capacity to remove CH_3 groups from C-4 and C-14 is missing. Both ^{14}C cholesterol and [4-^{14}C] 7-dehydrocholesterol are taken up from axenic media by *T. aceti* (Cole and Krusberg 1968), but neither *T. aceti* or *C. briggsae* liberated [14]CO_2 from [4-^{14}C] cholesterol (Rothstein 1968), though *C. briggsae* did form 7-dehydrocholesterol from cholesterol. *T. aceti* dealkylates plant sterols, e.g. β-sitosterol and fucosterol, to animal sterols, because triparanol succinate blocks this step when added to culture media with the accumulation of demosterol (Cole and Krusberg 1968).

An ecdysone-like material has been detected in *P. redivivus* and *A. avenae* by Dennis (1977*a*), but Dennis (1977*b*) was unable to show any effect of β-ecdysone on amino acid incorporations by *P. redivivus* polysomes *in vitro*. In insects the hormone β-ecdysone has such an effect. β-ecdysone was reported from *Ascaris suum* (Horn, Wilkie, and Thomson 1974).

5.3. Protein and amino acid metabolism

C. briggsae requires 10 essential amino acids in its diet (see Section 4.5) but nonetheless incorporates ^{14}C from [2-^{14}C] acetate, [1-^{14}C] glucose, and [2-^{14}C] glycine, when added to axenic culture media, into both the essential amino acids: threonine, tyrosine, valine, leucine, isoleucine, histidine, lysine, and arginine; and the non-essential amino acids: glutamic, aspartic, alanine, glycine, and serine (Rothstein and Tomlinson 1961, 1962).

The incorporation of ^{14}C from ^{14}C glucose and [2-^{14}C] acetate into several non-essential amino acids *in vitro*, has also been reported by the tylenchids *Ditylenchus triformis* and *Meloidogyne* sp., but more significant was the reported labelling of trytophan by *Meloidogyne*, an essential amino acid (Myers and Krusberg 1965). The incorporation of ^{14}C into 'essential' and 'non-essential' amino acids was similarly observed with *Aphelenchoides rutgersi* (Aphelenchida) by Balasubramanian and Myers

(1971). With $[1-^{14}C]$ acetate: glutamic, aspartic, proline, glycine, serine, cysteic acid, alanine, tyrosine, arginine, leucine + isoleucine, lysine, phenylalanine, threonine, and valine became radioactively labelled. One must conclude that the substitution of ^{14}C for non-labelled carbon occurs without total synthesis, or that the essential amino acids can be formed from metabolic precursors, but too slowly to maintain growth. The latter explanation seems less probable because other animals are thought to be unable to synthesize at least nine of the 'essential' amino acids.

Relatively little work has been done on amino acid metabolism, though Rothstein and his co-workers in the papers cited have outlined some probable pathways in *C. briggsae*. The transamination of pyruvate to alanine has been demonstrated in homogenates of *Aphelenchoides ritzemabosi*, with glutamate or aspartate as NH_2 donors (Miller and Roberts 1964) and glutamic oxaloacetate and glutamate pyruvate transaminases have been demonstrated in *Pelodera strongyloides* (Scott and Whittaker 1970). Folic acid (pteroylglutamic acid) is required by *C. briggsae* for the degradation of histidine, because in folic acid deficient media, population growth is restricted and formimino-L-glutamic acid accumulates (Lu *et al.* 1974). Cyanocobalamin (vitamin B_{12}) and folic acid are required in culture media for the conversion of homocysteine to methionine (an 'essential' S-amino acid) by *C. briggsae* (Lu *et al.* 1976).

A puzzling aspect of amino acid metabolism in nematodes is the excretion of amino acids. When *C. briggsae* was cultured in media containing ^{14}C-labelled metabolites, labelled amino acids and other compounds were excreted (Rothstein 1963). With $[2-^{14}C]$ acetate-labelled glutamate, aspartate, alanine, serine, and glycine were excreted. The excretion of labelled asparagine and glutamine has been reported from $[^{14}C]$ acetate, but nitrogen compounds are not the only compounds excreted (Rothstein 1965). When *C. briggsae* was cultured in media with $[2-^{14}C]$ acetate, labelled trehalose, glucose, and less glycerol were excreted. When *C. briggsae* was given labelled acetate in water instead of culture medium there was a great increase in glycerol. *T. aceti* behaved similarly, while with *P. redivivus* the principal labelled compounds were sugars in water and culture medium. *P. redivivus* also excretes a lot of amino acid nitrogen, as well as protein, when incubated in saline (Wright 1975a), the latter probably from the faeces. It is difficult to see amino acid excretion as a means of nitrogenous excretion, or explain the significance of the other products. Tylenchida excrete a variety of organic compounds (Myers and Krusberg 1965). *D. triformis* excreted amino acids, secondary amines, asparagine, aldehydes, and organic acids into solution *in vitro*. The larvae of *Meloidogyne incognita* excreted amino acids and sugars (Wang and Bergeson 1978). Rothstein (1965) concluded from his observations on the labelling of amino acids that, when a variety of different labelled metabolites were added to cultures of *C. briggsae*, there must be a continuous metabolic exchange of metabolites between the medium and the nematode.

In contrast to high levels of ^{14}C-labelled metabolites excreted by *T. aceti* in Rothstein's experiments, Zeelon, Gershon, and Gershon (1973) found it difficult to label the macromolecules of *T. aceti* by adding ^{14}C-, ^{3}H-, or ^{35}S-labelled precursors to the culture medium. Incorporation was poor and measurements of turnover by 'pusle label and cold chase' experiments proved impractical. They concluded that the synthesis of macromolecules depended on a large internal pool of nucleotides and amino acids, while assimilation of precursors was poor. This conflicts with Rothstein's view that there is a ready exchange between the medium and the body. Pasternak and Samoiloff (1970) did not find similar difficulties in labelling proteins in *Panagrellus silusiae* with [^{3}H] leucine.

Chromatin has been extracted and purified from *C. elegans* (Vanfleteren, Neirynck, and Huylebroeck 1979). Histones were analysed by gel electrophoresis, revealing four of the five main groups: H1, H2a, H2b, H4, with H3 probably combined with H2a. Eighteen non-histones were detected including myosin, paramyosin, actin, and tropomyosin.

Muscle proteins have also been extracted and purified from *C. elegans* and their properties studied (reviewed by Zengel and Epstein 1980). F-actin, myosin, paramyosin, and tropomyosin, constituting the contractile apparatus in nematodes, as in other animals, have been characterized (Harris and Epstein 1977; Harris, Tso, and Epstein 1977). *In vitro*, under appropriate conditions, these proteins form unique paracrystalline filaments and show characteristic interactions and enzymic properties. Myosin heavy chains possess a Mg^+ dependent ATPase, which is regulated by the calcium concentration, and, in the absence of ATP, myosin binds to F-actin. F-actin (MW 42 000) and tropomyosin (MW 40 000) show structural differences from those of other animals, but demonstrate typical biochemical reactions associated with the calcium regulation of contraction.

There are three different myosins in *C. elegans*: one with a heavy chain of MW 206 000 localized in the oesophageal muscles, and two other structurally distinct myosins occurring together in the body-wall muscle cells, both with heavy chains of MW 210 000. Studies of muscle-defective mutants have demonstrated the differences between these two myosins (see Section 2.6 for references). The myosin molecule has two heavy chains and four light chains (two MW 18 000 and two 16 000). Paramyosin (MW 98 000) has been localized in the body wall and oesophageal muscles (Waterston, Epstein, and Brenner 1974; Harris and Epstein 1977; Zengel and Epstein 1980). *In vitro* it can form filaments around a core of myosin.

TABLE 5.3

The DNA base composition of several nematodes

| Species | % (guanine + cytosine) | | | |
	From buoyant density	From thermal denaturation	From base composition	Reference
Panagrellus silusiae	44	44	—	Behme and Pasternak (1969)
P. redivivus	44	44	—	Behme and Pasternak (1969)
Caenorhabditis briggsae	36	39	—	Behme and Pasternak (1969)
C. elegans	36	36	35.5*	Sulston and Brenner (1974)
Rhabditis anomala	42	42	—	Behme and Pasternak (1969)
Turbatrix aceti	40	not done	—	Behme and Pasternak (1969)
Aphelenchoides rutgersi	—	—	41.3	Thirungnam and Myers (1974)
Ascaris lumbricoides	—	39	—	Kaulenas and Fairbairn (1968)
Ascaris lumbricoides	37	—	—	Bielka, Schultz, and Böttger (1968)

* A small fraction hybridizing with *r*RNA has 51 per cent guanine–cytosine.

5.4 Nitrogenous excretion

Ammonia is the major nitrogenous excretory product of nematodes (Wright and Newall 1976) and its accumulation may limit nematode survival in culture. The rate of excretion is a function of body weight in high, probably a consequence of the general inverse relationship between size and metabolic rate in animals (Wright and Newall 1976; and discussion in Section 3.6). *D. triformis* lost 9–18 mg N/100 g wet weight/24 hours when incubated *in vitro*, equivalent to 10 µmoles N/g wet weight/24 hours, 23–41 per cent as amino acids and 27–44 per cent as ammonia (Myers and Krusberg 1965; Wright and Newall 1976). *P. redivivus* excreted much more nitrogen *in vitro*, 213 µmoles/g wet weight/24 hours (Wright 1975*a*; Wright and Newall 1976), of which 25–35 per cent was ammonia, 6–7 per cent urea, as well as amino acid and protein nitrogen; the protein perhaps as faeces.

The excretion of urea rather than ammonia is unusual in aquatic animals, and energetically expensive, but may minimize the toxicity of ammonia. All the enzymes of the ornithine urea cycle (carbamyl phosphate synthetase ornithine transcarbamylase, argininosuccinic acid synthetase, argininosuccinic acid lyase, and arginase) and evidence for a functional cycle and the excretion of urea were reported in *P. redivivus* (Wright 1975*b*). No evidence was found in this species for purine degredation to ammonia or urea, because xanthine oxidase was low and urease, uricase, and allantoinase were not found, but purines probably contribute relatively little to nitrogen excretion. In *Pelodera strongyloides* uricase, xanthine oxidase, uric acid, and urea were detected (Scott and Wittaker 1970), but uric acid and urea were not found in the medium. Ammonia is excreted by *C. briggsae*, and Rothstein (1963) did not find urea, uric acid, creatinine, nor allantoin excreted. Amino acid excretion has already been discussed (see Section 5.3).

5.5. Nucleic acids and nucleic acid metabolism; cyclic nucleotides

The base composition of DNA has been considered a clue to phylogenetic affinity between major taxa, and is usually expressed as guanine + cytosine as a percentage of the total bases. The values for several nematodes are shown in Table 5.3. Base composition has been reported for *Aphelenchoides rutgersi* by Thirungnam and Myers (1974) as

	Percentage of molar composition				
	Guanine	Adenine	Cytosine	Thymine	Uracil
DNA	22.5	29.5	18.8	29.3	—
RNA	31.4	22.8	22.9	—	23.0

The ratio of guanine to cytosine was 1.2:1, but such a departure from theoretical equivalence has been reported in other animals. DNA accounted for 0.9 per cent of dry weight, RNA 2.6 per cent.

In *C. elegans*, Sulston and Brenner (1974) estimate the haploid genome contains 8×10^7 base pairs, 83 per cent of which is unique, smaller than any other known from animals and only about 20 times that of *E. coli*. DNA has been extracted from *C. elegans* by Vanfleteren *et al.* (1979) in the course of isolating histones from chromatin (see Section 5.3), with average ratios of: histone/DNA 1:1.01, non-histone/DNA 1:10.72, and RNA to DNA of 1:0.10.

As already noted Zeelon *et al.* (1973) had difficulty in labelling the ribonucleic acids of *T. aceti* by adding [^{3}H] uridine to culture media. Their best results were with [^{14}C] formate (which also labels amino acids).

Pasternak and Samoiloff (1970), however, successfully labelled proteins, DNA and RNA in *Panagrellus silusiae* with [^{3}H] leucine, [^{3}H] thymidine, and [^{3}H] uridine. With *A. rutgersi* Thirungnam and Myers (1974) compared the rates of incorporation of [^{14}C$_2$] glycine into purines with that of [5-^{3}H] orotic acid into pyrimidines. The higher rate of incorporation of [5-^{3}H] orotic acid suggested purine synthesis may be relatively inadequate, supporting their nutritional studies (see Section 4.5).

Cyclic nucleotides, which typically act as 'the second intracellular messenger' when animal cell membrane receptors bind circulating hormones, have been found in *P. redivivus* and it seems likely that they fulfil this function in nematodes (Willett 1980). Cyclic AMP (adenosine 3', 5'-cyclic monophosphate) was reported from axenic culture (Willett and Rahim 1978*a*) as well as cyclic GMP (guanosine 3', 5'-cyclic monophosphate) (Willett and Rahim 1978*b*). The enzymes necessary for synthesis, cyclic diesterases, are also present and there is circumstantial evidence for the presence of the phosphodiesterases required for their degradation (Willett 1980). The concentraton of cAMP varies with age, rising from 0.1 fmol per individual in early larvae to 2.0 in post-reproductive adults (Willett, Rahim, and Bollinger 1978), and adenylate cyclase levels similarly rise sharply with age. cGMP rose from 0.1 fmol per individual larva to 22 in the adult and then fell to 2.0 in senescent adults.

5.6. Intermediary metabolism and bioenergetics

Much more is known about the respiratory metabolism of parasitic nematodes than about that of free-living nematodes and so it is not known whether the differences between parasitic helminths and higher animals are specifically adaptations to parasitism, or characteristic of nematodes and other helminths, both parasitic and non-parasitic. Many parasitic nematodes inhabiting oxygen-deficient environments within their hosts are facultative anaerobes, which utilize oxygen when available, but are predominantly anaerobic. However, at some stage of their life-cycle they require oxygen, in *Ascaris* for example, during embryonic development which takes place outside the host. It may be that anaerobic respiration evolved as an adaptation to life in the alimentary canal of a host, or it may be that the capacity to respire anaerobically evolved first in free-living nematodes inhabiting oxygen-deficient aquatic sediments, thereby becoming pre-adapted to parasitism. It would be interesting to know whether nematodes from such habitats can respire anaerobically.

It will be inappropriate to discuss metabolism of parasitic nematodes at length here, but it is necessary to draw attention to some of the salient features in order to see what is known about free-living nematodes in perspective. The respiratory metabolism of parasitic nematodes has been reviewed by Barrett (1976*a*, *b*), and Bryant (1975, 1978). Most work has been done on *Ascaris*, which differs from higher animals in glycolysis, the tricarboxylic acid cycle (TCA-cycle), and the electron transport system. I shall presume a general familiarity with these systems in mammals. A characteristic feature of *Ascaris* metabolism is the secretion of reduced organic acids.

Glycolysis in most animals follows the Embden–Meyerhoff pathway, converting glucose anaerobically to two molecules of pyruvate with the coupled phosphorylation of ADP. Some enzymes of the Embden–Meyerhoff glycolytic pathway have been reported from *Ditylenchus triformis* and *D. dipsaci* (Krusberg 1960) and *P. redivivus* and *T. aceti* (see Table 5.2; Ells 1969; Barrett *et al.* 1971; Zeelon *et al.* 1973). All the enzymes of this pathway have been found in parasitic nematodes, but the pathway operates differently.

In *Ascaris* and other parasitic helminths, the pathway branches at phosphoenolpyruvate (PEP), one branch following the classical vertebrate pathway in which PEP is dephosphorylated to pyruvate with the formation of ATP. In the alternative branch, PEP is carboxylated to oxaloacetate with the phosphorylation of GDP (see Fig. 5.1). In *Ascaris* the alternative pathway predominates. The relative importance of the two pathways is dependent on the kinetics of the two enzymes involved, i.e. phosphoenol-pyruvate carboxykinase and pyruvate kinase. In helminths enzyme kinetics

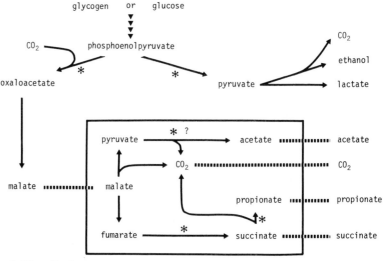

* sites of ATP synthesis

Fig. 5.1.   Energy metabolism in helminths. Alternative pathways for the catabolism of carbohydrate and generation of ATP with their characteristic end products. Steps shown within the box take place within the mitochondrion; steps outside the box in the cytosol. Not shown, but necessary for catabolism to be maintained is a balance in oxidation and reduction of NAD, both in the mitochondrion and the cytosol, so that reduced NADH is available for reductive reactions. (I am grateful to Dr C. Behme for this diagram.)

favours the fixation of carbon dioxide by PEP, whereas in vertebrates, kinetics favours the decarboxylation of oxaloacetate to PEP. The importance of CO_2 fixation and the role of the alternate pathway has been reviewed by Bryant (1975). The oxaloacetate formed is reduced to malate.

The significance of malate formed from PEP is that it enters the mitochondrion where further anaerobic metabolism generates ATP, and gives rise to the volatile acid end-products of anaerobic metabolism (see Fig. 5.1). The anaerobic metabolism of *Ascaris* mitochondria has been reviewed by Rew and Saz (1974). Part of the malate is converted to pyruvate and then acetate by successive decarboxylations, while the remaining malate gives rise to fumarate, which is reduced to succinate with the phosphorylation of ADP. These reactions involve compensating oxidations and reductions of NAD, so that NADH is available to maintain glycolysis in the cytosol anaerobically, and for the conversion of acetate, pyruvate, and succinate to volatile fatty acids. In vertebrate tissues, where pyruvate is the end-product of the classical pathway, the pyruvate is reduced to lactate under anaerobic conditions, with the reoxidation of NADH. Thus the nature of the excreted end-product is a clue to respiratory pathway operating. Some nematodes are lactate excretors under anaerobic and aerobic conditions.

Different parasitic nematodes vary in the relative importance of the alternative pathways from PEP and their end-products: lactate, acetate, or volatile fatty acids. It is not known which pathway is taken in free-living nematodes, though the excretory products liberated by *Caernorhabditis* sp and *A. avenae* (Cooper and Van Gundy 1971) point to a significant use of the pyruvate lactate pathway under anaerobic conditions. Lactate accumulated in the tissue but was not excreted, while ethanol excretion accounted for a high proportion of the glycogen utilized anaerobically, presumably by the reduction of pyruvate. [^{14}C] ethanol, when added under aerobic conditions, was oxidized to [^{14}C]O_2. Succinate was liberated aerobically and anaerobically.

The pathway from oxaloacetate to succinate represents a reversal of the TCA-cycle. The complete cyle has very low activity in adult *Ascaris* tissues, but may be more significant in other parasites. The isolation of TCA-cycle enzymes and TCA acids suggest it is also active in *Ditylenchus triformis* (Krusberg 1960; Castillo and Krusberg 1971). Studies of ^{14}C-labelled metabolites in *C. briggsae* (e.g. Rothstein 1965) and the oxidation of TCA substrates by a mitochondrial preparation from *T. aceti* (Rothstein, Nicholls, and Nicholls 1970) suggest active TCA-cycles in these two species. Enzymes from the complete TCA-cycle have been reported from *P. redivivus* and *T. aceti* (Table 5.4; Ells and Read 1961; Barrett *et al.* 1971; Castillo and Krusberg 1971). The TCA-cycle completes the oxidation of carbohydrates and lipids, which enter the cycle as acetyl co-enzyme A.

TABLE 5.4

The specific activities of various enzymes in homogenates of two species of bacteria-feeding nematodes (from Barrett, Ward, and Fairbairn 1971)

Enzyme	Specific activity in n moles/min/mg protein 30 °C; mean ± standard error			
	Panagrellus redivivus		*Turbatrix aceti*	
β-oxidation				
Acyl-CoA synthetase (long chain)	13.1	± 1.3	17.8	± 0.6
Acyl-CoA dehydrogenase	16	± 1.3	7.3	± 0.2
Enoyl-CoA hydratase	4631	± 118	2182	± 127
3-hydroxyacyl-CoA dehydrogenase	870	± 32	110	± 12
Acetyl-CoA acyltransferase	150	± 16	24.3	± 3.8
TCA and glyoxylate				
Citrate synthase	103.8	± 5.6	160	± 13.5
Aconitate hydratase				
citrate to *cis*-aconitate	162	± 4.9	58.3	± 2.8
isocitrate to *cis*-aconitate	488.5	± 18.3	162.75	± 18.5
Isocitrate dehydrogenase (NADP)	153.8	± 7.1	31.2	± 1.02
Succinic dehydrogenase	68.3	± 4.3	65	± 4.8
Fumerate hydratase	706.7	± 86	238	± 18.3
Malate dehydrogenase	2812	± 195	770	± 60
Isocitrate lyase	82	± 8.5	37	± 3.8
Malate synthase	76	± 6	47	± 2
Glycolytic and glyconeogenic				
Pyruvate kinase	24.3	± 1.6	92	± 5.8
Phosphoenolpyruvate carboxykinase	36.25	± 3.75	188	± 16.7
Lactate dehydrogenase	52.3	± 1.0	8.0	± 3.5
Fructose-1 6-diphosphatase	16.2	± 1.6	22.6	± 1.56

Means are for 5–10 assays; protein by Lowry method; buffer: 0.25 M sucrose, 0.005 M tris-HCl, pH 7.5.

Pyruvate enters by reductive decarboxylation forming acetyle CoA, which is finally converted to CO_2 and water. The TCA-cycle and the associated electron transport systems with coupled phosphorylations occur within the mitochondria.

Lipids and glycogen are the food reserves of parasitic and free-living nematodes. Lipid enters the TCA-cycle as acetyl co-enzyme A from the β-oxidation of fatty acids. All the enzymes necessary for the β-oxidation of fatty acids are present in *T. aceti* and *P. redivivus* (Barrett *et al.* 1971), and both can utilize uniformly labelled [^{14}C] palmitate (see Table 5.4). The utilization of endogenous lipids and glycogen, when starved under aerobic conditions, by *Caernorhabditis* sp. and *Aphelenchus avenae* has been discussed already (Table 5.2, Cooper and Van Gundy 1970). Anaerobically *A. avenae* uses up its lipid reserves, which are replaced at the expense of glycogen when oxygen is restored, probably accounting for its greater tolerance of anaerobiosis than *Caenorhabditis*.

It is ironic that although *Ascaris* was one of the first animals in which cytochromes were described by Keilin in the 1920s, the nature and role of cytochromes in *Ascaris* remains controversial after many subsequent studies. Barrett (1976*a*), after reviewing the history of studies of nematode cytochromes, concludes that all the animal-parasitic nematodes can respire aerobically when oxygen is present, but that their cytochrome systems differ from mammalian cytochromes. Cytochromes in higher animals are associated with the transfer of electrons from reduced compounds to oxygen, coupled with oxidative phosphorylations. Cheah (1976*a*) believes that *Ascaris suum* muscle mitochondria contain a functional branched cytochrome respiratory chain with two terminal oxidases, i.e. cytochromes o and a_3. The classical mammalian chain, cytochromes b, c_1, c, a, and a_3 is present, with the bifurcation at the b level. Cheah (1976*b*) has described the properties of cytochrome b-560 from *Ascaris* mitochondria. Under anaerobic conditions, fumarate replaces oxygen as the terminal electron acceptor, with the reduction of fumarate to succinate coupled to the phosphorylation of ADP, perhaps involving a b-type cytochrome (Köhler 1980).

Little is known about the electron transport system in the aerobic respiration of free-living nematodes, or the coupling of electron transport to oxidative phosphorylation, or whether fumarate is reduced to succinate under anaerobic conditions with the coupled phosphorylation. Spectrophotometric evidence of cytochromes c and a have been found in *D. triformis*, where there was evidence of a complete TCA-cycle (Krusberg 1960). Histochemical evidence for cytochrome oxidase has been found in *Pratylenchus screbneri*, *C. briggsae*, and *P. redivivus* (Deubert and Zuckerman 1968).

Cytochromes can be identified spectrophotometrically from their difference spectra, following their reduction with strong chemical reducing agents, or by potential TCA-enzyme substrates. Difference spectra in the presence of inhibitors, such as cyanide or carbon monoxide, are often of value. Rothstein *et al.* (1970) obtained mitochondrial preparations from *T. aceti* which, when reduced by NADH or succinate, gave difference spectra characteristic of cytochromes a (600 nm), b (560 nm), and c plus c_1 (550 nm). An additional haemoprotein was present in considerable excess of the cytochromes, which with dithionite reduction superimposed on substrate reduction, gave a peak at 568 nm with a 'shoulder' at 535 nm. Apparently the same haemoprotein, when exposed to carbon monoxide after reduction, gave peaks at 568, 573, and 420 nm.

Stimulation of oxygen uptake by the mitochondrial preparation on the addition of TCA-enzyme substrates, and the effect of inhibitors, suggest cytochrome a_3 is a terminal oxidase in *T. aceti*. With succinate as substrate, plus ADP, a P:O ratio of 1.3–1.5 was obtained for phosphorylation. Incidentally, mitochondria with unusual cristae have been described from electron microscope observations on *T. aceti* (Zuckerman, Kisiel, and Himmelhoch 1973).

With living *C. briggsae*, 10^{-4} M cyanide inhibits oxygen uptake by 70 per cent and exposure to a gas mixture of 90 per cent CO + 5 per cent O_2 inhibits uptake by 55 per cent (Bryant *et al.* 1967). Growth and reproduction cease, but resume on removing the inhibitors, while oxygen consumption can be partially restored. Inhibition by carbon monoxide is partially reversed by light. The effect of the inhibitors on oxygen uptake is consistent with the properties of cytochrome a_3 as a terminal oxidase, while the residual respiration may indicate a cyanide and carbon monoxide insensitive cytochrome b. Bryant *et al.* (1967) observed spectrophotometric evidence of a b-type cytochrome in low-temperature (77 K°) difference spectra from intact and particulate preparations from homogenates of *C. briggsae*. A reduced difference spectrum revealed a peak at 557 nm with small peaks corresponding to the other cytochromes.

Thus, in summary, the few free-living nematodes investigated appear to have an active full TCA-cycle, coupled to an electron transport system culminating in cytochrome a_3. However, they may possess, in addition to the classical chain of cytochromes b, c_1, c, a + a_3, other reducible haemoproteins, which may be b-type cytochromes. Moreover, they show varying capacities to tolerate anaerobic conditions, though evidence of the PEP, malate, fumarate, succinate pathway, characteristic of anaerobic parasite respiration is so far lacking. The end-products of carbohydrate metabolism in *Caenorhabditis* and *A. avenae* suggest the operation of the classical pathway from phosphoenolpyruvate to lactate and ethanol, though the evidence is not unequivocal.

Haemoglobins are present in many parasitic nematodes, where their role in metabolism is controversial. Haemoglobin has been found in the oesophagus of a marine free-living nematode, *Enoplus brevis*, by Atkinson (1975). Microspectrophotometric measurements of oxyhaemoglobin gave peaks at 577.6, 543.6, and 421.7 nm and at 555.2, 532.2 nm when deoxygenated. This haemoglobin is important in oxygen transport in *E. brevis* Atkinson (1977). ADP has been measured in *Anguina tritici*, *P. redivivus*, and the eggs of the *Meloidogyne incognita*, by luciferase bioluminesce. In *P. redivivus* one larva had about 1.22 ng ATP (Spurr 1976). ATP levels associated with cryptobiosis in *A. avenae* are referred to in Section 3.9.

The glyoxylate cycle is important in plants and micro-organisms, but was not known from animals until found in *C. briggsae* by Rothstein and Mayoh (1964; 1965; 1966). The complete cycle was demonstrated in *C. briggsae*, *T. aceti*, *P. redivivus* and *Rhabditis anomala* by Rothstein and Mayoh (1965, 1966). The cycle requires several TCA-cycle enzymes and isocitrate lyase and malate synthetase to convert 2-carbon compounds to

4-carbon compounds (see Fig. 5.2). Its importance lies in providing a mechanism by which acetate derived from the β-oxidation of fatty acids can be used for protein and carbohydrate synthesis. Acetate stimulates the growth of *T. aceti* (see Chapter 4). In *T. aceti* catalase, isocitrate lyase, D-amino oxidase, and α-hydroxyacid oxidase were separated from

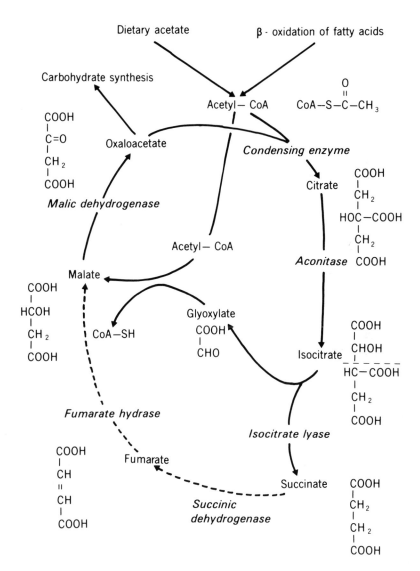

FIG. 5.2. The glyoxylate cycle through which fatty acids may contribute to carbohydrate synthesis.

mitochondrial cell fractions in 'peroxisomes' on a sucrose gradient by Aueron and Rothstein (1974).

In *C. elegans* the activity of the cycle is high in the eggs, falls during development, and rises again in reproductive adults (Patel and McFadden 1978*a*). In *C. elegans* isocitrate lyase, malate synthetase, and catalase are located in cellular organelles termed glyoxysomes (Patel and McFadden 1977), but Rubin and Trelease (1976) associated isocitrate lyase and malate synthetase with mitochondria not glyoxysomes in *Ascaris* larvae. The properties of isocitrate lyase from *C. elegans* and *A. suum* have been reported by Reiss and Rothstein (1974), Colonna and McFadden (1975), and Patel and McFadden (1978*b*).

5.7. Digestive enzymes

Digestive glands are well developed in many nematodes which frequently possess more than one type of oesophageal gland. Oesophageal gland excretions may be released into the lumen of the oesophagus, or the buccal cavity, or extruded through the mouth or stylet. The nature of their enzymic secretions can usually only be guessed at from the nature of their food. However, the enzymes secreted have been characterized in some species, mostly plant-feeding nematodes. The identification of a hydrolytic enzyme within nematode tissues does not show whether this enzyme plays any part in digestion, though a pattern of isozymes, identifiable in homogenates by gel electrophoresis, may be of taxonomic interest (see Section 9.4). With most enzymes, chemical identification of the enzyme needs to be combined with histochemical localization of its origin.

Very little is known about digestion in bacteria-feeding nematodes. Jennings and Deutch (1975) showed that β-glucuronidase occurs in the intestine of adults, but not larval, *Monhystera denticulata* (marine: Monhysterida). It is apparently secreted by cells of the cardia and continues to act within intestinal cells following assimilation. It probably digests bacterial cell walls. These authors also demonstrate arylamidase (leucine amino peptidase) in the seminal recepticle of females of *M. denticulata* and *Chromadorina germanica* (Chromadorida), where it may digest old spermatozoa. Some histochemical observations have been made on the digestion of blood by the omnivorous marine nematode *Pontonema vulgaris* (Enoplida) by Jennings and Colam (1970). Esterases have generally been associated with the nervous system and have been discussed in Section 3.4, but also hydrolyse fatty acids. They have been reported from intestinal tissues of *P. vulgaris* (Jennings and Colam 1970) and Dorylaimida (Lee 1964). In the predatory freshwater dorylamid, *Actinolaimus hintoni*, Lee (1964) found cholinesterase and non-specific esterases around the stylet, perhaps used in feeding. Many different phosphatases and esterases were separated from homogenates of *P. redivivus* and *D. triformis* by gel electrophoresis (Benton and Myers 1966) but their function is unknown.

The enzymes that attack plant cell walls, pectinases and cellulases, are important in Tylenchida. Deubert and Rohde (1971) critically discuss our knowledge of these enzymes in nematodes at that time. Cellulases have been assayed viscometrically, or by the release of reducing substances, but usually not from natural cellulose. Cellulose activity has been reported from homogenates or suspending media from many species (e.g. Tracey 1958; Krusberg 1960; Goffart and Heiling 1962; Morgan and McAllan 1962; Dropkin 1963; Bird, Downton, and Hawker 1975), but with one exception (Myers 1965) not from the few bacteria-feeding Rhabditida tested. Similarly pectinases, which include a number of different enzymes specific for different pectins, have been found in many Tylenchida, but not apparently in other nematodes (Deubert and Rohde 1971). Chitinases have been found in three *Ditylenchus* species, but not *T. aceti* (Tracey 1958). Chitin occurs in fungal cell walls on which *Ditylenchus* feeds; also in nematode egg shells. Amylases in contrast occur in Tylenchida and Rhabditida (e.g. Goffart and Heiling 1962; Morgan and McAllan 1962; Myers 1965; Gysels 1968; Krusberg 1960). Proteolytic enzymes are present in homogenates of a number of nematodes from both orders, including pepsins, trypsins, and exopeptidoses (Miller and Jenkins 1964; Gysels 1968), and trypsin has been reported from *Meloidogyne incognita* (Dasgupta and Gunguly 1975).

5.8. Biochemistry of development

The generalization that in nematodes nuclear division is largely completed during embryonic development, apart from the gonads, has perhaps been overemphasized. In *C. elegans* the number of non-gonadial nuclei nearly doubles between the hatching of the first larval stage and reproductive adults. Nonnenmacher-Godet and Dougherty (1964) reported the incorporation of [^{3}H-] thymidine into non-dividing nuclei of the intestine of *C. briggsae*, as well as the germinal zone of the ovary, by autoradiography. Tritiated thymidine was progressively lost in the presence of non-tritiated thymidine. The authors suggested an exchange of methyl radicals or gene amplification, but continued nuclear division of intestinal cells, as in *C. elegans* (Sulston and Horvitz 1977) could produce similar results. However, polytene or polyploidy has been demonstrated in *Ascaris* tissues (Swartz, Henry, and Floyd 1967).

Macromolecular synthesis associated with growth can be blocked by appropriate inhibitors. In *Panagrellus silusiae* growth in length can be inhibited by adding actinomycin, puromycin, and phleomycin to culture media (Pasternak and Samoiloff 1970), and reduced by hydroxyurea and phenetyl alcohol but not by nalidixic acid. Nucleic acid and protein synthesis, measured by the incorporation of [³H] uridine and [³H] leucine, respectively, were blocked by actinomycin D, actidione, and hydroxyurea to varying degrees. Actinomycin D, which blocks transcription of DNA into RNA, inhibits growth and stops the development of the reproductive system of *P. silusiae* (Boroditsky and Samoiloff 1973). Actidione, a less specific blocking agent, also inhibited growth and gonad development. Interestingly hydroxyurea, believed to act by inhibiting DNA synthesis, disorganized gonad development, with little effect on growth. The use of inhibitors to study sexual attraction has been discussed in Section 3.5.

Acrylamide gel electrophoresis of homogenates from age-synchronized cultures of *P. silusiae* shows changes in the pattern of proteins at different stages of development (Chow and Pasternak 1969). There are also differences in the isozymes of lactic dehydrogenase, malate dehydrogenase, and esterase at different stages. Collagen synthesis, as measured by the incorporation of [³H] proline continues throughout development with peaks preceding moults, apparently without re-utilization of protein from moulted cuticle (Leushner and Pasternak 1975).

Aminopterin, a folic-acid antagonist, inhibits growth in axenic cultures of *C. briggsae* and *T. aceti* (Dougherty and Hansen 1957; Kisiel, Nelson, and Zuckerman 1972). Low doses affect formation of the vagina. Aminopterin, 5-fluorodeoxyuridine, and hydroxyurea all inhibit the growth, reduce the longevity, and prevent the maturation of *C. briggsae* and *T. aceti* (Kisiel, Nelson, and Zuckerman 1972). The effects of aminopterin treatment of *C. briggsae* depend to some degree on the stage of growth when applied and the nutritional value of an axenic medium (Vanfleteren and Avau 1977). In a poor medium for growth the effect of removing it may be less lasting than in a good medium. The authors believe aminopterin effectively causes thymine deficiency in larval stages.

6. Developmental biology

6.1. Introduction

NEMATODES have interested developmental biologists because they have proved the most suitable animals with which to investigate certain problems in developmental biology. Late in the nineteenth century, when biologists were interested in the role of chromosomes in development, and in the concept of the separateness of the germ plasm from generation to generation, the eggs of the horse roundworm *Parascaris equorum* were favoured subjects for investigation. The sequence of cell divisions, the continuity of chromosomes through successive divisions, and the developmental fates of embryonic cells were followed in great detail. Recently, renewed interest has been shown in the sequence of cell divisions, or cell lineages, by which the fertilized egg gives rise to the adult nematode. *C. elegans* with its short generation time, relatively few cells, and ease of maintenance in the laboratory, has proved ideally suited for this purpose and its embryonic development has been studied in great detail.

Nematode development is highly determinate. In *C. elegans* the sequence of cell divisions, migrations, and differentiations are essentially invariant. Recent work with experimentally induced mutations which collectively affect every aspect of development, is advancing our knowledge of the genetic basis for development, as will be discussed in the next chapter. Short generation times and strictly limited cell division have made several nematode species favoured subjects for research on ageing.

The less well known development of several other metazoan phyla probably closely resembles nematode development, for example, that of Rotifera, Gastrotricha, and Acanthocephala. They contrast strikingly with other phyla, for example Platyhelminthes, in which the development of the individual is much less rigidly determined. In Platyhelminthes, developmental potentialities of undifferentiated cells, which persist in the adult, are wide. Associated with this difference in the degree to which development follows a strictly determined pattern, there are differences in the diversity of life-cycles and methods of reproduction to be found within phyla. Asexual reproduction is not found in nematodes, though they do show interesting variations in sexual reproduction (discussed in the next chapter), and the life-cycle is highly stereotyped. The Platyhelminthes display a great diversity in their life-cycles and methods of both asexual and sexual reproduction. Nematodes show virtually no ability to regenerate damaged organs following injury, whereas many Platyhelminthes show great capacities for regeneration.

6.2. Embryonic development

Embryonic development can be observed in the egg through the transparent egg shell. At the close of the nineteenth century, important observations were made on the development of the horse roundworm *Parascaris equorum* (Zur Strassen 1896; Boveri 1899) which demonstrated that cell divisions followed a rigid sequence with the developmental fates of each cell highly predictable. The fertilized egg of *P. equorum* has only four chromosomes whose unusual behaviour made it possible to trace the potential germ cell line back to the fertilized egg, because only the chromosomes of the germ-cell line do not undergo chromosome diminution. In chromosome diminution the chromosomes fragment at mitosis, the middle euchromatic regions giving rise to subunits which will behave like separate chromosomes, while the heterochromatic club-shaped ends disintegrate completely (Boveri 1887).

The first free-living nematode in which embryonic development was described in comparable detail was *Turbatrix aceti* (Pai 1927). Since then the embyronic development of many nematodes has been described, though generally in less detail, and it is clear that embryonic development follows a very similar course in all nematodes. The first cleavage divisions of ascarids are unusual in giving a T-shaped early embryo instead of an oval shape. Apart from Dorylaimida, the embryology of Adenophorea is little known, but recent work has shown some differences in other Adenophorea in early cleavage from that of Secernentea (Milutina 1981; Malakhov 1981). Most observations have been made on the intact egg, though Chuang (1962), used histological sections to follow changes in cellular structure after gastrulation in *Pelodera* (= *Rhabditis*) *teres*. Recently, Normarski differential interference contrast optical microscopy has made it possible to follow cellular proliferation and rearrangements within the intact eggs of *C. elegans* (Deppe, Schierenberg, Cole, Krieg, Schmitt, Yoder, and von Ehrenstein 1978; von Ehrenstein and Schierenberg 1980; and Fig. 6.1). Electron microscopy has shown the fine structural detail of cellular differentiation in *C. elegans* and, with the aid of computer analysis of electron micrographs, has been used to construct models of cell structure which completely tally with Normarski images (Krieg, Cole, Deppe, Schierenberg, Schmitt, Yoder, and von Ehrenstein 1978).

In the nematode-egg, the first cleavage division gives rise to an AB cell and a P_1 cell, the second, four cells giving rise to progenitors of the ectoderm A and B cells; the endoderm-plus-mesoderm, EMSt cell; and propagative cell, P_2-cell (gonads plus some posterior ectoderm and proctodaeum; see Fig. 6.1). EMSt stands for endoderm, mesoderm, and stomodaeum (called S_2 for somatic by Pai). Pai interpreted the first four divisions as successive separations of somatic cells from the future germ cells, following Boveri's terminology, in which the P-cell can be traced back to the ovum, Po. Building on the classical work of Boveri and Pai, but interpreting embryonic development in the light of more recent concepts of cellular differentiation, one can give a very detailed picture of development in *C. elegans* (Deppe *et al.* 1978; von Ehrenstein, Schierenberg, and Miwa 1979).

An important concept is that of stem cells. The early cleavage divisions are asymmetric giving rise to a stem cell and a smaller P-cell, according to the scheme shown in Fig. 6.2. These authors recognize six stem cells, AB, MSt, E, C, D, and P_4 in *C. elegans*, giving rise to primary ectoderm,

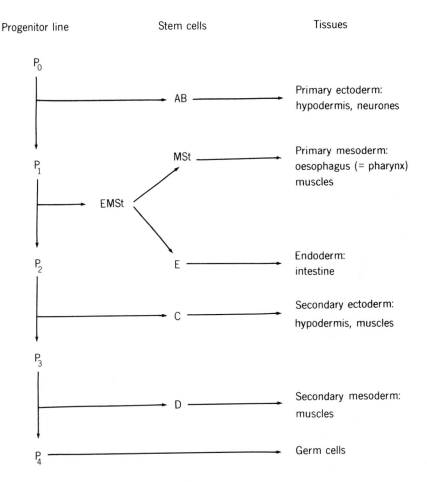

FIG. 6.1. A diagrammatic representation of embryonic development of *C. elegans* showing the origin of stem cells and the fate of their descendants based on the work of von Ehrenstein and Schierenberg (1980).

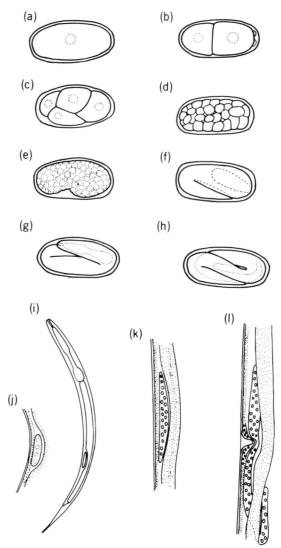

FIG. 6.2.   The embryonic development of *C. elegans* (a) to (h): (a) zygote, (b) first cleavage, AB to the right with two polar bodies, P_1 to the left; (c) second cleavage; (d) post-gastrulation; (e) lima bean stage; (f) plum stage; (g) pretzel stage; (h) fully developed larva. Several intermediate stages are recognized, a comma and tadpole stage intermediate between the lima bean and the plum stage and a loop stage between the plum and pretzel stages. Gonoduct development in *C. elegans* during larval stages, (j) to (l): (j) and (i) four-cell stage in first larval stage; (k) proliferation of gonad primordium; and (l) the reflexing of the tips of the gonads and the intrusion of hypodermal cells to form the vulva in the late larval hermaphrodite. (Original drawings of development, with descriptive names from von Ehrenstein and Schierenberg (1980).)

primary mesoderm, endoderm, secondary ectoderm, secondary mesoderm, and gonad primordia, respectively. Stem cells are cells whose
dependants arise by synchronous and symmetric cell divisions. Rates of
division differ between different cell lines, controlled by endogenous
clocks. Waves of mitotic divisions, initiated by the AB-cell line and
generally proceeding from anterior to posterior, pass over the embryo, but
because of the differing rates in the different lines these overlap. Cell lines
maintain their position in the division sequence even when displaced by
gastrulation. The E-cell line can be distinguished from its eighth division by
autofluorescence. The numbers of cell divisions and their orientation are
invariant, and differentiation and migration are determined by a cell's
place in the sequence of divisions. Division of P_4, which comes to lie
mid-ventrally within the body, is delayed until later in larval life, to then
gives rise to both the germ cells and most of the gonoducts.

The stomadaeum, arising from the EMSt cell, is mesodermal, giving rise
to the oesophageal musculature, but is anomalous in being lined by cuticle,
apparently secreted by muscle cells, though differing in structure from that
secreted by the ectoderm. However, ectodermal cells derived from AB
contribute to the oesophagus as neuroblasts and epithelial cells (Chuang
1962; von Ehrenstein and Schierenberg 1980). The stomodaeum becomes
visible as a solid core of cells with less dense cytoplasm early in
organogenesis. The proctodaeum, also usually considered ectodermal, is
also lined by cuticle.

Most neuroblasts, derived from the ectoderm, become internal at
gastrulation, others indistinguishable from hypodermal cells enter later,
mainly by internal/external division. Neither nerve fibres nor synapses
have developed at the lima bean stage, when gastrulation has been
completed, and their identification as neuroblasts rests on following their
subsequent differentiation.

Developing nematode embryos pass through a characteristic series of
forms, termed by von Ehrenstein *et al.* (1979) ball, lima, bean, comam,
tadpole, plum, loop, and pretzel (see Fig. 6.2). I am not sure of the
descriptive appropriateness of plum. The ball, about 500 cells in *C.
elegans*, marks the completion of rapid proliferation, and the initiation of
organogenesis, largely complete by the pretzel stage. The pretzel is a fully
formed larva coiled up within the egg. In *C. elegans*, the pretzel stage is
mobile, indicative of nerve and muscle differentiation; and has a fully
formed oesophagus (with valves) and intestine (with microvilli) (Kreig *et
al.* 1978).

Electron micrographs of *C. elegans* (Kreig *et al.* 1978) at about the
294-cell stage shows P_4 surrounded by a layer of endodermal cells and an
outer layer of ectoderm. At the lima bean stage (540 cells) a group of
endodermal cells is surrounded by a single layer of mesodermal cells, the

primordial body wall musculature, within a single layer of ectodermal cells. The early embryo, unlike the hatched larva, is not bilaterally symmetrical. Bilateral symmetry is achieved by groups of cells rotating from anterior–posterior to left–right (MSt and D), or migration to left–right (C) (von Ehrenstein and Schierenberg 1980).

In *C. elegans* much of embryonic development is probably determined in the unfertilized egg. Temperature-sensitive mutants, which prevent the development of the eggs, have been described by Hirsch (1979), who also describes experimental crosses with which one can distinguish between those genes which must be expressed in the egg before fertilization, i.e. maternal-effect genes, from those whose expression is necessary after fertilization. Much of cell division is dependent on maternal-effect genes, presumably requiring *m*RNA or proteins from these genes to be present in the fertilized egg. Significantly, nucleoli do not appear in embryonic nuclei until the 26-cell stage. Hirsch (1979) analysed 24 developmentally defective *zyg*-phenotypes which arose from mutation of maternal-effect genes. Von Ehrenstein *et al.* (1979) described 11 temperature-sensitive mutant genes affecting embryonic development (9 genes in chromosome III and 3 on other chromosomes). With seven of these, expression of the maternal gene is sufficient for embryogenesis and in only one was zygote expression essential.

Von Ehrenstein and Schierenberg (1980) list many temperature-sensitive mutant genes, with some acting specifically at every step in the development of *C. elegans*. Of particular interest are mutant genes which affect the duration of cell division cycles, suggesting this may play a key role in development. With *emb*-5, for example, which leads to premature death of the embryo, the E-cell line divides too rapidly. One mutant, *lin*-5 controls timing of chromosome replication independently of cell division, resulting in giant polyploid cells.

6.3. Cell division, cell lineages, and regulation in embryonic and post-embryonic development

Very detailed observations on the sequence of cell divisions by which the fertilized egg of *C. elegans* gives rise to the embryonic organs and tissues, i.e. cell lineages, have been made and given in detail by von Ehrenstein and Schierenberg (1980). Observations of cell lineages and cell rearrangements have been followed through into the adult with cell division continuing in some tissues throughout larval development. Cell lineages have been followed through into the adult of *C. elegans* by Sulston and Horvitz (1977), who followed by Normarski optical microscopy the fates of nuclei in individual nematodes which had been anaesthetized by chilling or 1-phenoxy-2-propanol. At hatching, the first-stage larva has 550 nuclei, which increase to 810 somatic nuclei in the hermaphrodite and 970 in the male. The pattern of cell division is almost identical in all individuals of the same sex.

Most post-embryonic divisions occur in the hypodermis, gonads, accessory sexual organs, intestine, muscles, and nervous system. The anterior nervous system and oesophagus appear complete at hatching. The fates of some cells, as determined by observations of their nuclei, vary according to the sex of the individual. A mesodermal cell (M) generates either the male accessory sex musculature or the vulval and uterine muscles according to the sex of the nematode. Post-embryonic nuclear multiplication was also reported in *P. redivivus*, *T. aceti*, *Aphelenchoides blastophthorus*, and *Longidorus macrosoma* by Sulston and Horvitz (1977). In *P. silusiae* the numbers of nuclei increase throughout development in muscles, nerves, hypodermis and intestine (Sin and Pasternak 1970). Intestinal nuclei, unlike those of other tissues, increase their DNA content (by microspectrophotometry) to many times that of haploid spermatozoa, equivalent to 16–32 ploid.

Most cell lineages in *C. elegans* are invariant in all individuals, though some very minor variations have been observed. Development requires some cells to migrate to new locations. The interesting question is whether the fates of cells are strictly controlled by the sequence of cell divisions, i.e. their lineages, or whether differentiation is regulated by position effects and the interactions between tissues. Sulston and Horvitz (1977) concluded from their observations that essentially invariant cell lineages generate a fixed number of cell progeny of rigidly determined fates. This conclusion was supported by observing the effects of destroying cells during development by laser beam. This method of examining the capacity to regulate development was reported more fully in Kimble, Sulston, and White (1979). Very limited regulative changes of several kinds following the death of a cell were observed. In general, specialized cells were replaced by a neighbouring cell according to a hierarchy of potential fates. The possibility of replacement was restricted to small groups of cells (equivalence groups) of apparently related ancestry and was reflected in imperfect development of adult structures.

6.4. Post-embryonic growth

There are four moults in the life-cycle of nematodes, separating four larval stages, the final moult preceding full sexual maturity. The gonads develop progressively during larval life and are usually present as a compact body with four nuclei when the first larval stage of either sex hatches, lying mid-ventral to the intestine (see Fig. 6.1). Post-embryonic development has been described in: *C. elegans* (Hirsch, Oppenheim, and Klass 1976;

Sulston and Horvitz 1977), *Panagrellus silusiae* (Boroditsky and Samoiloff 1974), *Pratylenchus* sp (Roman and Hirschmann 1969), *Ditylenchus dipsaci* (Hirschmann 1962), *Diploscapter coronata* (Hechler 1968) amongst others. The two central nuclei, derived from P_4, give rise to the primordial germ cells, the remaining two to the terminal gonoduct epithelium. In didelphic females, or hermaphrodites, the division of the epithelial nuclei separates the germ cell nuclei so that they become the germ cells of two opposite gonads. In monodelphic species one gonad may degenerate leaving a post-vulval sac. In other monodelphic species, and in males, both germ cell nuclei remain adjacent as the epithelial nuclei divide, giving rise to a single gonad.

Cells from the ventral hypodermal chord contribute to the terminal gonoducts and vulva. In *C. elegans* the cell lineages of accessory sexual organs are similar in males and hermaphrodites until late in development, but may, as with the sex muscles, differentiate differently according to sex, or may, as in the development of male copulatory bursa, undergo additional cell divisions in the male (Sulston and Horvitz 1977). Additional ventral chord divisions give rise to 14 male-specific neurones. A linker cell in the male dies to join the vas deferens to the cloaca and is the analogue of an anchor cell which joins the uterus to the vulva in the hermaphrodite. Mesodermal cells form accessory muscles in both sexes. A mesoblast (M) gives genital muscles in males; uterus and vulva in hermaphrodites.

The presence of a cuticle, which is moulted at intervals, does not impose stepwise growth like that typical of arthropods, and substantial growth can occur without moulting, especially after the last moult. However, discontinuities in growth and activity are associated with moulting. *C. elegans* becomes inactive 1–2 hours before moulting (at 20°), termed lethargus, with the cessation of oesophageal pumping its most obvious feature (Cassada and Russell 1975). Byerly *et al.* (1976) used an optical method for rapidly sizing and counting nematodes to study the growth of *C. elegans*. Growth in the length of age-synchronized populations followed a smooth sigmoid curve from hatching to adult, maximal about the fourth moult. However, small perturbations were detectable in the curve, associated with slight changes in body proportions associated with the times of moulting. Wilson (1976) studied a synchronous population, measuring the distribution of lengths and concluded that 'steps' in the growth corresponding to mid-moult were detectable in *C. elegans* and *P. redivivus*.

In parasitic nematodes growth may cease at one larval stage, usually the third, pending infection of a host. In free-living nematodes, under favourable conditions, growth may proceed continuously from hatching to sexual maturity. In many Rhabditidae, when conditions become unfavourable, some larvae develop differently, to form an environmentally more resistant resting stage, a dauer larva. Many nematodes can become quiescent, enter diapause or cryptobiosis, under unfavourable conditions (see Section 3.9). Dauer larvae develop in cultures of *Caenorhabditis*. With *C. briggsae* they may develop in rich axenic media, where the second moult of the dauer larvae is delayed. The dauer larva, longer than non-dauer second-stage larvae, has a closed mouth, anus, and intestine, and is more tolerant of high temperatures (Yarwood and Hansen 1969). It resumes a normal life-cycle when transferred to fresh media. With *C. elegans* second-stage dauer larvae accumulate in old monoxenic cultures, from which they can be selectively isolated by having greater resistance to sodium dodecylsulfate than other stages (Cassada and Russell 1975). The dauer larvae are relatively inactive, without pharyngeal pumping, but tend to 'stand on their tails' on agar surfaces (wink larvae) and are relatively resistant to toxic chemicals, osmotic stress, desiccation, and heat. The dauer larva is thinner than other larvae, has a closed intestine, does not feed, and has the basal layer of the interchordal cuticle striated. Starvation induces dauer larval formation and dauer larvae resume development in a new medium (without any change in the normal sex ratio).

Cassada and Russell (1975) described a temperature-sensitive mutant giving very high proportions of dauer larvae when the temperature is raised to 25 °C. Two types of mutants affecting dauer larval formation are known: those which cause dauer formation when food is abundant, constitutive mutants (affecting perhaps 10 genes); and those which prevent dauer formation on starvation, defective mutants (perhaps affecting 30–60 genes). Some dauer defective mutants also show defective chemotactic responses and ultrastructural abnormalities in their amphid sensilla (Riddle 1980).

6.5. Ageing and senility

An interest in gerontology, especially the problems arising from human senility, has stimulated work on nematodes, because they have offered the most suitable experimental animals for certain lines of investigation. An underlying hypothesis is that some common mechanisms of ageing operate in all animals. The advantages offered by *Caenorhabditis*, *Panagrellus*, and *Turbatrix* are short life spans, simple axenic culture of large numbers of age-synchronized individuals, and highly determinate development, so that old individuals contain old cells. Many signs of advancing age have already been established in these genera (Zuckerman 1974; Zuckerman and Himmelhoch 1980). In *C. elegans* these include decreasing locomotory activity, though pharyngeal activity is sustained (Croll, Smith, and Zuckerman 1977), increased osmotic fragility, increased specific gravity, and increased formaldehyde sensitivity (Zuckerman, Himmelhoch, Nelson, Epstein, and Kisiel 1971). These may be related to changes observed in the cuticle (Zuckerman *et al*. 1971; Zuckerman, Himmelhoch, and Kisiel 1973*a, b*) and/or phospholipids. Zuckerman and Himmelhoch (1980) argue that ageing in nematodes is fundamentally similar to that in higher animals.

The intracellular deposition of lipoidal pigment has attracted interest because of the association of pigment deposition with age in other animals. It has been reported in *C. briggsae* (Epstein and Gershon 1972), *P. redivivus* (Buecher and Hansen, reported by Zuckerman and Himmelhoch 1980), and *T. aceti* (Kisiel and Zuckerman 1974). Senility in *C. elegans* is associated with lower fecundity, primarily through affecting the oocyte (Béguet and Brun 1972). However, after a number of generations selectively raised from the progeny of old parents fecundity returns to normal levels (Béguet 1972). In *P. redivivus* the levels of cyclic nucleotides change (Willett, Rahim, and Bollinger 1978; see Section 5.14).

Inhibitors have also been used to stop reproduction in all three experimental species, especially 5-fluorodeoxyuridine, hydroxyurea, and aminopterin. Rothstein (1980), who has reviewed the use of inhibitors in detail, concludes that the many potential side-effects of metabolic inhibitors makes the interpretation of results less certain than when mechanical means of age synchronization are employed. Rothstein and his co-workers have perfected efficient means of synchronizing *T. aceti*, using repeated separation on sieves (Hieb and Rothstein 1975; Rothstein 1980).

Polyacrylamide gel electrophoresis has revealed different patterns in the isozymes of lactic dehydrogenase and malic dehydrogenase with age in *P. silusiae* (Chow and Pasternak 1969). Changes in enzymes have been studied more thoroughly in *T. aceti*. Hydrolase activity increases; α-amylase and malic dehydrogenase decreases (Erlanger and Gershon 1970). With isocitrate lyase and fructose-1, 6-diphosphate aldolase an increasing proportion of the molecules are enzymically inactive as the nematode ages (Gershon and Gershon 1970; Zeelon *et al*. 1973). Decreases in specific activity per mole have been reported in isocitrate lyase, triosephosphate isomerase, and phosphoglycerate kinase (Reiss and Rothstein 1974, 1975). Differences have also been found in t-RNA (Reitz and Sanadi 1972).

In both mammals and *T. aceti*, purified enzymes from older animals differed from the same enzymes from younger animals. In *T. aceti* five enzymes have been compared, isocitrate lyase, phosphoglycerate kinase, aldolase, enolase, and triose isomerase, but only the last did not show changes (Rothstein 1980). Rothstein's conclusions, following a detailed review of the evidence, much of it from his own collaborative work, supports the hypothesis that these changes are primarily due to a slowing down in the rate of turnover of the enzymes, with the accumulation of altered enzymes, and not to errors of transcription or translation.

7. Genetics and cytogenetics

7.1. Introduction

THE use of nematodes in genetic research is expanding at an unprecedented rate, with most of the research on a single species, *Caenorhabditis elegans*. The widespread and growing interest in this species was inspired by Dr Sydney Brenner and his co-workers at the Medical Research Council Laboratory of Molecular Biology in Cambridge, England, who selected this species for studies of developmental genetics in 1965. They needed an animal suitable for genetic analysis, with a simple nervous system, so that the genetic control of the development of the nervous system could be investigated. The production and genetic analysis of many different mutants, affecting morphological, behavioural, physiological, biochemical, and cytological characters have made this nematode valuable for investigating many fundamental problems in developmental biology. An important development has been the establishment of the *Caenorhabditis* Genetics Center at the University of Missouri (Division of Biological Sciences, Columbia, Mo, USA 65211).

The rapidly growing literature on the biology of *C. elegans* in English has tended to overshadow the numerous contributions in French on the genetics, cytogenetics, and developmental biology of *C. elegans* and other nematodes. A pioneer in this field for many years was Professor Victor Nigon (e.g. Nigon 1949) who built on the work of earlier European nematologists (Maupas 1900; Krüger 1913; Hertwig 1920; amongst others). The late Professor J. Brun and his co-workers at the University of Lyons in France have been very active in work on nematode genetics.

Two strains of *C. elegans* are being used for experimental genetics. *C. elegans* Bristol strain, which was chosen by Brenner for his research is the most widely used and was obtained from Dr E. C. Dougherty's laboratory. I originally collected this species in 1956 from a sample of commercial mushroom compost, from the vicinity of Bristol in England, which had been provided for a short course in agricultural nematology by Dr L. N. Staniland. The French biologists tend to use *C. elegans* Bergerac, which was collected by Nigon from garden soil at Bergerac in France (Fatt 1964). I put both strains into axenic culture for Dr Dougherty (Nicholas *et al.* 1959), but whether the cultures now in use in France come from the same or separate isolates, I do not know.

C. elegans cultures normally contain a great preponderance of self-fertilizing hermaphrodites, with very rare males. One mutant population designated *him* (high incidence of males) is available to facilitate genetic analysis by experimental crosses. A generation time of only 3½ days at 20 °C with 250–350 progeny is advantageous for genetic experiments. Self-fertilization ensures a high degree of homozygosity and that recessive mutations will not be apparent in the first-generation progeny (F_1), but will segregate as clones in one-quarter of the F_2 progeny without any risk of unwanted crosses. Clones can be stored frozen in liquid N_2.

7.2. Gametogenesis

The organization of the nematode gonad in some ways facilitates observations on gametogenesis by light microscopy. The successive stages are arranged in a linear sequence along the length of the gonad and individual chromosomes can be seen in mitosis and meiosis in stained preparations of the whole organ (by Feulgen's technique or with orcein; Triantaphyllou 1979). On the other hand the organ is very small, as are the chromosomes, while in spermatogenesis the numerous superimposed spermatocytes obscure one another. With oogenesis, developing occytes, fertilized oocytes, and eggs in early cleavage become arranged in single file. Oogenesis has been described in great detail in *C. elegans* by light microscopy (Nigon and Brun 1955; Hirsch, Oppenheim, and Klass 1977) and electron microscopy (Abirached 1974; Abirached and Brun 1975), but also in varying detail in many other Rhabditida and Tylenchida (e.g. Triantaphyllou 1971*b*, 1981; Triantaphyllou and Hirschmann 1962, 1966, 1967; Roman and Triantaphyllou 1969), Aphelenchida (Triantaphyllou and Fisher 1976), and Dorylaimida (e.g. Dalmasso and Younes 1969, 1970).

The germinal zone of the gonad lies at the tip of the organ and in young adults in both sexes contains oogonia, undergoing mitotic division. In *C. elegans* about 1300 oogonial nuclei cluster around an axial acellular cytoplasmic rachis, derived from four cap cells (Hirsch *et al.* 1977) at the extreme tip, which extends as far as the point of flexure. Such a rachis is found in other nematodes and may be a general feature of nematode oogenesis. Its presence during spermatogenesis, which in self-fertilizing hermophroditic *C. elegans* precedes oogenesis in the same gonad (in the larval stage), has not been clarified. There are breaks in plasma membranes incompletely separating oogonial cells from the rachis.

Oogonia pass from mitotic proliferation to meiotic division in a very short region in which their densely staining chromosomes are in synapsis. Synapsis, the pairing of homologous chromosomes, initiates prophase of the first meiotic division (Meiosis I), in which the diploid primary gametocyte will give rise to secondary haploid gametocytes. In nematodes the earliest stages of prophase: leptotene, zygotene, and diplotene, are difficult to distinguish, but have been described in *C. elegans* (Nigon and

Brun 1955; see Fig. 7.1). As in other animals, both oocyte meiotic divisions are highly unequal, giving rising to a single ovum and two or three polar bodies (Polocytes). However, completion of the first meiotic division, culminating in the separation of homologous chromosomes and the second meiotic division (Meiosis II), in which the two chromatids constituting each chromosome separate, are completed only after the spermatozoon has entered the ovum.

Following the zone of synapsis, there follows a relatively long zone with the chromosomes in pachytene. In oogenesis, oocytes grow at the expense of the rachis as the nucleus enlarges, while the compact chromosomes which initially stain deeply, gradually become thread-like and more diffuse until they cannot be clearly defined. At pachytene synaptonemal complexes form. Paired homologous chromosomes are separated along their length by a constant gap (about 100 nm in electron-micrographs) across which they are connected by numerous parallel fibrils (Abirrached 1974).

As oocytes pass the region of gonad flexure they enter the zone of

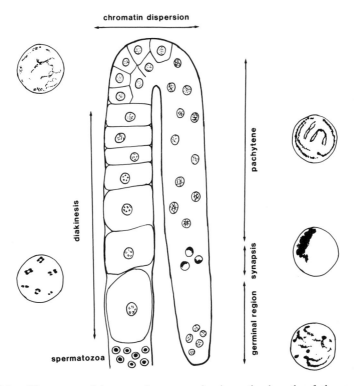

FIG. 7.1. The sequential stages in oogenesis along the length of the ovary of *Caenorhabditis elegans*, based on drawings and photomicrographs published by Nigon and Brun (1955).

diakinesis. Oocytes continue to grow, filling the whole organ in single file, while the chromosomes condense again. Nematode chromosomes in diakinesis show some characteristic features. As the chromosomes condense, the bead-like chromosomes coalesce into short rods. V-shaped chromosomes associate in homologous pairs, bivalents, characteristically end to end, appearing as a cross or ring. The separation of chromatids in each bivalent gives the appearance of tetrads.

Triantaphyllou and Fisher (1976), after studying *Aphelenchus avenae*, believe that chromosomal behaviour in Aphelenchids and *Caenorhabditis* can be interpreted as follows. The chromosomes split into two chromatids early in prophase I, which remain side by side (as dyads) until separating at anaphase II. When the homologous chromosomes pair they therefore appear as tetrads. Initially, in diplotene, the bivalents are associated along

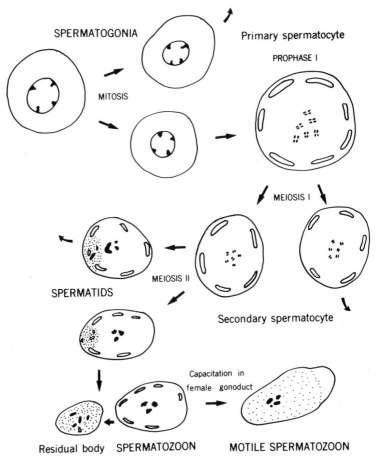

Fig. 7.2.   Successive stages in spermatogenesis based on Shepherd (1981).

their central (euchromatic) regions where chiasmata presumably occur as in other animals. At diakinesis the chiasmata become terminalized so that the homologous chromosomes become associated end to end, though in *C. elegans* this may not extend to the heterochromatic tips, giving the appearance of a cross. The chromosomes are probably acrocentric and sub-acrocentric, hence the end-to-end association. Alternatively, their 'sideways' movement at Metaphase II may indicate a diffuse kinetochore.

Spermatogenesis follows a similar pattern to oogenesis in Rhabditida (Nigon 1949; Nigon and Brun 1955) and Aphelenchida (Triantaphyllou and Fisher 1976; Shepherd 1981), but is usually much more difficult to follow, because meiotic divisions are equal and the spermatocytes do not grow as much as the ovum. The spermatids produced by a meiotic division transform into spermatozoa in the male gonoduct, with further maturation after entering the female gonoduct. Some confusion in nomenclature has been clarified by Shepherd (1981) (see Fig. 7.2). Nematode spermatozoa are non-flagellate, lack an acrosome, and neither nucleoli nor nuclear membranes reform after Meiosis II. Their chromatin occurs in dense aggregates. These changes have been described in *Panagrellus silusiae*, *Rhabditis pellio*, and *Aphelenchoides blastophthorus* (Pasternak and Samoiloff 1972; Beams and Sekhon 1972; Shepherd and Clark 1976*a*, respectively), and several parasitic nematodes. *A. blastophthorus* has the advantage that the developing spermatozoa are arranged in single file.

Bundles of fibrils and membranous organelles, of unknown function, appear in developing nematode spermatozoa. The closely aligned fibrils form within the spermatocytes, initially closely associated with double-membraned organelles. In the spermatids the membranous organelles proliferate, becoming cisternae of complex shape, sometimes with internal microvilli. Transition from spermatid to spermatozoon occurs with the cutting off from the spermatozoon of a residual body, taking with it cytoplasm, ribosomes, and endoplasmic reticulum. The cisternae of the membranous organelles open to the surface of the spermatozoon while in the male gonoduct or after being deposited in the female. Perhaps they are functionally equivalent to the acrosome of other animals.

In the female gonoduct, further reorganization of the organelles occurs and pseudopodia develop at one pole of the spermatozoon which becomes amoeboid. In *A. blastophthorus* the cells lining the vas deferens of the male gonoduct are secretory, their membranes becoming closely apposed obscuring the duct lumen. Copulation and insemination in *P. redivivus*, and its effects on the life-cycles of both sexes, have been described by Duggal (1978*b*, *c*). Two different secretions from the male gonoduct are introduced with the spermatozoa into the female duct during copulation, in which between 20 and 100 spermatozoa are transferred. The male may produce 2000–3000 spermatozoa in total.

Fertilization occurs within the female gonoduct. Entrance of a spermatozoon into the ovum stimulates the completion of Meiosis I (without reformation of the nuclear membrane), followed immediately by Meiosis II. Each division gives rise to a peripheral polar body (polocyte), into which one set of anaphase chromosomes migrate. The first polar body may divide again, giving a total of three polar bodies. The membranes of the male and female pronuclei disperse, and both male and female chromosomes contribute to the metaphase of the first cleavage division of the zygote nucleus. In nematodes, the chromosomes derived from the male and female may be distinguished by their position at metaphase of the first cleavage.

7.3 Methods of reproduction, sex determination, and sex ratios

Nematodes reproduce exclusively by eggs, without alternative asexual or vegetative mechanisms, but display interesting variations on the normal bisexual mode of reproduction. We can recognize six basic patterns:

    (i) bisexual amphimictic (cross-fertilized eggs);
   (ii) bisexual pseudogamous parthenogenetic;
  (iii) hermaphroditic, usually automictic, but occasionally amphimictic;
  (iv) hermaphroditic pseudogamous parthenogenetic;
   (v) thelytokous mitotic parthenogenetic;
  (vi) thelytokous meiotic parthenogenetic.

Maupas (1900) made a detailed study of the reproductive habits of terrestrial free-living nematodes but credits Schneider with first describing a form of self-fertilizing hermaphroditism common in nematodes, as well as observing parthenogenesis in nematodes. Much research has been devoted to unravelling the cytogenetic basis for these different modes of reproduction.

These diverse modes of sexual reproduction facilitate the appearance of polyploid or aneuploid races in nematodes, a phenomenon much commoner in plants than in most other animals. The changes in the karyotype (chromosome complement) have been important in the evolution of nematodes and have been studied extensively, especially by Professor A. C. Triantaphyllou and his co-workers at North Carolina State University in the United States.

Most nematodes are bisexual and cross-fertilization (or amphimixis) is the commonest method of reproduction. However, many Rhabditida, some Tylenchida, and at least one Dorylaimida (*Mononchus* (= *Clarkus*?) *papillatus*; Steiner and Heinly 1922) are protandrous hermaphrodites, in which the hermaphrodite is morphologically female; but produces first spermatozoa then oocytes in the same gonad. The oocytes are fertilized by spermatozoa stored in the gonad, i.e. automixis, but not infrequently non-viable unfertilized eggs continue to be laid after the spermatozoa have been used up. Some species show repeated phases of spermatogenesis and oogenesis in the same gonad, e.g. *Rhabditis gurneyi* (Potts 1910).

Automictic species often show very unequal sex ratios, while amphimictic species generally show ratios near unity. In such automictic species, the rare males are capable of copulation with hermaphroditic 'females' and fertilization of the eggs. The sex ratio, expressed by the Andric index (percentage of males in the population), is often between 0.01 and 2 in predominantly automictic Rhabditida, many examples of which are cited by Maupas (1900) and Nigon (1949). Nigon (1949), who made a detailed study of sex ratios and sex determination, showed that it was genetically determined. A line of *C. elegans* giving high proportions of males has been made available for genetic research (see Section 7.1), compared with a normal index of about 0.2.

Parthenogenesis is common in Rhabditida, Tylenchida, and Dorylaimida. In mitotic parthenogenesis, maturation divisions of the oocyte are abnormal, chromosomes do not pair at synapsis, and only a single mitotic division occurs resulting in an egg and one polar body, both with the somatic chromosome number (Hechler 1968; Triantaphyllou 1971a). In meiotic parthenogenesis two meiotic maturation divisions occur giving rise to an oocyte with the haploid chromosome number. Restoration of the diploid chromosome number occurs in various ways, for example by fusion of the second polar body nucleus with the oocyte nucleus, failure of the chromosomes to separate at Meiosis II, or by endomitosis during the first cleavage division of the egg (Triantaphyllou 1966, 1970, 1971a, b, 1981; Triantaphyllou and Fisher 1976; B'Chir 1979).

In parthenogenesis, the development of the egg is initiated by something other than fusion of the spermatozoon nucleus with that of the oocyte. One way, pseudogamy, is through penetration of the oocyte by the spermatozoon without fusion of the male and female pronuclei. Pseudogamy was first described from a hermaphroditic rhabditid, *Rhabditis aberrans* by Krüger (1913), and later from other bisexual and hermaphroditic species (Hertwig 1920, 1922; Belar 1923; Goodchild and Irwin 1971). It may be equivalent to mitotic or meiotic parthenogenesis in different species (Triantaphyllou 1971a).

Sex determination appears generally to be of the XX, XO type with the male the heterogametic sex, as in *Pelodera strongyloides* and *Caernorhabditis elegans* (Nigon 1949) (see Table 7.1). In some species, with equal chromosome numbers in both sexes, determination is probably of the XX, XY type, as in *Aphelenchus avenae* (Triantaphyllou and Fisher 1976) and *Mesorhabditis belari*, where there is evidence that the male is the heterogametic sex (Nigon 1949). The sex chromosomes of Rhabditida can

TABLE 7.1
Diploid chromosome numbers of Rhabditida

Species	Female or hermaphro-dite	Male	Reproduction	Reference
Rhabditidae				
R. aberrans	18	17	automictic	Krüger (1913)
R. maupasi	14		automictic	Hertwig (1922)
R. anomala	20		pseudogamous	Goodchild and Irwin (1971)
R. pellio	14	13	amphimictic	Goodchild and Irwin (1971); Hertwig (1920)
Caenorhabditis dolichura*	12	11	automictic	Honda (1925); Nigon (1949)
C. elegans*	12	11	automictic	Honda (1925)
C. briggsae*	12	11	automictic	Nigon and Dougherty (1949)
Mesorhabditis belari*	20	20	amphimictic and pseudogamous	Nigon (1949)
Pelodera strongyloides*	22	21	amphimictic and pseudogamous	Nigon (1949)
Diplogasteridae				
Diploscapter coronata	2		parthogenetic	Hechler (1968)
Fictor anchicoprophaga	12		amphimictic	Pillai and Taylor (1968)
Paroigolaimella bernensis	12		amphimictic	Pillai and Taylor (1968)
Mononchoides changi	14	13	amphimictic	Hechler (1970)
Neodiplogaster pinicola	14	13	amphimictic	Hechler (1972)
Cylindrocorpidae				
Cylindrocorpus longistoma	12	13	amphimictic	Chin and Taylor (1969b)
C. curzie	14	13	amphimictic	Chin and Taylor (1969b)
Cephalobidae				
Panagrolaimus rigidus	8	8	amphimictic	Nigon (1949)
Panagrellus redivoides	10	9	amphimictic	Hechler (1972)

* Previously referred to genus *Rhabditis*.

sometimes be recognized by their retarded disjunction during Meiosis I (Nigon 1949; Hechler 1972). Nematode chromosomes are usually very small short rods, so that in species with large numbers of chromosomes, it is difficult to recognize individual chromosomes, as for example in *Anguina tritici* with $2n = 38$ (Triantaphyllou and Hirschmann 1966).

The tendency for the sex chromosomes to separate late at Meiosis I of spermatogenesis can account for the rare occurrence of males in predominately hermaphroditic species. Failure of disjunction can lead to a small proportion of the secondary spermatocytes lacking an X chromosome, giving rise to some fertilized eggs of XO constitution. This has been observed in *Caenorhabditis dolichura* (Nigon 1949) and probably occurs in other species (Krüger 1913). Support for this hypothesis comes from observations of an experimentally produced tetraploid line of *C. elegans* by Nigon (1951). Sex chromosomes tended to be lost over many generations so that fully tetraploid automictic individuals ($2n = 24$) produced a small proportion of automictic progeny with $2n = 23$ and rare males with $2n = 22$. Nigon was able to calculate the probability of loss of one or both sex chromosomes at meiotic division.

Facultative parthenogenesis occurs in *Diplenteron potohikus* (Diplogasteridae), a bacteria-feeding nematode from sand dunes. A proportion of the eggs develop into functional males in crowded cultures, but not when isolated. A proportion of the eggs can be induced to develop as males by an extract from culture media or by various indole compounds (Clark 1978). Environmental factors also influence the development of males in facultatively parthenogenetic Heteroderidae, as in several species of *Meloidogyne* (Triantaphyllou 1971a). In *Aphlenchus avenae* parthenogenetic and amphimictic races with $n = 8$ (in both sexes of the latter) are known, as well as a parthenogenetic race with $n = 9$ in which males are produced by high temperatures (Hansen *et al.* 1973; Triantaphyllou and Fisher 1976).

7.4 The karyotype

The chromosome complement can be determined by staining the metaphase chromosomes during meiosis or early cleavage, using the Feulgen technique or orcein (Triantaphyllou 1979). The karyotypes of many plant-parasitic Tylenchida, Aphelenchida, and Dorylaimida have been determined because they throw light on the origins and phylogenetic histories of many important parasites of plants. The karyotypes of many other non-parasitic terrestrial nematodes are also known, but have not been so systematically investigated (see Table 7.1). The karyotypes of many Tylenchida have been described and illustrated by Triantaphyllou and co-workers (Triantaphyllou and Hirschmann 1962, 1966, 1967; Roman and Triantaphyllou 1969; Triantaphyllou and Fisher 1976). In *Aphelenchoides*, species with $2n = 3, 6$, and 8 are known. *A. tuzeti* with $2n = 3$ is obligately parthenogenetic, diploidy being restored by failure to form a polar body at Meiosis I (B'Chir and Dalmasso 1979). Plant-feeding species of *Xiphinema* with $n = 5$ or 10, and *Longidorus* with $n = 7$, include amphimictic and mitotic parthenogenetic species (Dalmasso and Younes 1969, 1970; Dalmasso 1975).

Within some tylenchid families, the evolution of plant parasitism has been associated with parthenogenetic reproduction and varying degrees of polyploidy. It is possible to construct convincing phylogenetic histories for these families from their karyotype, taxonomic affinities, and host plant specificities. The best examples come from the Heteroderidae (Triantaphyllou 1970, 1981). In *Heterodera* the basic haploid number was probably nine, i.e. $n = 9$, and for 13 amphimictic species, the haploid number is still nine. Mitotic parthenogenesis has apparently arisen more than once with diploid or triploid numbers from 24 to 34. Meiotic parthenogenesis may be an intermediate step with $n = 9$ to 13, as in *H. betulae*. In *Meloidogyne* the basic haploid number seems to be $n = 18$, as in one strictly amphimictic species and several combining amphimixis and parthenogenesis. Mitotic parthenogenesis is associated with diploid and triploid numbers from 36 to 54, while meiotic parthenogenesis is associated with aneuploidy which varies within a species from $n = 17$ to 15. During the pachytene stage of Meiosis I pairs of homologous chromosomes come together to form synaptonemal complexes, with one end attached to the nuclear envelope (as in other animals). The number of synaptonemal complexes is evidence of ploidy and has been studied in species of *Heterodera* and *Meloidogyne* (Goldstein and Triantaphyllou 1978a, b, 1979, 1980, 1981).

A similar evolutionary history can be suggested for *Pratylenchus*: with amphimitic species with $n = 5$, 6, or 7; a meiotic parthenogenetic species with $n = 6$; and mitotic parthenogenetic species with $2n$ or $3n = 25–36$ (Roman and Triantaphyllou 1969). Within the genus *Helicotylenchus* one amphimictic species has $n = 5$, while mitotic parthenogenetic species have somatic numbers from 30 to 40, suggesting high levels of ploidy, followed by the loss of several chromosomes (Triantaphyllou and Hirschmann 1967). Data on other Tylenchida is given by Triantaphyllou (1971a) in a review of nematode cytogenetics.

Mitotic parthenogenesis seems associated with high levels of ploidy and meiotic parthenogenesis with chromosomal instability. What may happen is illustrated by several free-living bacteria-feeding species of *Rhabditis* commonly associated with earthworms (Goodchild and Irwin 1971). *R. pellio* is strictly bisexual and *R. maupasi* usually hermaphroditic, both with $2n = 14$. *R. anomala*, with $2n = 20$, is a pseudogamous species, probably a triploid, formed through the addition of a haploid set through the aberrant fertilization of an unreduced diploid egg by a spermatozoon. Hertwig (1920) described a mutant of *R. pellio* reproducing by pseudogamous parthenogenes, a possible intermediate state.

Some plant-parasitic nematodes combine bisexual reproduction with parthenogenesis, depending on conditions. Parthenogenesis may, by permitting reproduction of isolated individuals, facilitate the exploitation by sedentary species of the presence of host plants, while retaining the genetic advantages of sexual reproduction at higher population densities. Perhaps this is why polyploidy is frequently found in Tylenchida. However, karyotypic variation within genera is not restricted to plant-parasitic genera. In *Seinura*, a predatory aphelenchid genus, the haploid number varies from three to six in bisexual species, showing varying degrees of amphimictic and automictic (hermaphroditic) reproduction, and corresponding differences in their sex ratios (Hechler 1963; Hechler and Taylor 1966a). In the Dorylaimida, the predatory family Monochidae includes amongst several genera: amphimictic species with $n = 8$; meiotic parthenogenetic species with $n = 5, 7, 8, 14$; and probable polyploid species with $n = 14$ and 28 (Cuany and Dalmasso 1974).

7.5. Genetic analysis

The genetic analysis of experimentally induced mutations of *C. elegans* has been fundamental to much of the extraordinarily productive research now being undertaken with this species. The essential methodology and basic genetics of *C. elegans* were described in an important paper by Brenner (1974), whose methods have generally been adopted by later workers. A more recent review of the genetics of this species and of the methods which have proved useful in studying its genetics has been given by Herman and Horvitz (1980).

Brenner (1974) exposed young adult self-fertilizing hermaphrodites for four hours at 20 °C to a chemical mutagen, ethyl methanesulfonate (EMS), which is believed to cause point mutations. About 1 in 1000 of the progeny had a visible mutation. This gives an estimated forward mutation rate of about 5×10^{-4} per gene. Both the spermatozoa and the ova may carry mutations, which through self-fertilization, give rise to heterozygous first-generation offspring (F_1). Recessive mutations will become apparent in 25 per cent of the second generation (F_2) progeny of self-fertilizing F_1s, producing homozygous clones of mutants. Brenner used monoxenic cultures with *Escherichia coli* on an agar.

Males are uncommon in cultures of *C. elegans* (0.3 per cent), but are required for genetic analysis. A line of *C. elegans* selected for giving rise to numerous males, *him* (high frequency males), is available for this purpose. The mutation responsible, which is an autosomal recessive reduces fecundity, but produces about 21–5 per cent male progeny. An autosomal dominant mutation producing a similar proportion of males and a proportion of sterile hermaphrodites has been described by Beguet (1978).

Earlier attempts to induce mutations in *C. elegans* by chemicals were reported by Nigon and Dougherty (1954), who had previously described a heat-induced dwarf mutation (Nigon and Dougherty 1949). More efficient methods used have been X-irradiation, ^{32}P-irradiation, and a variety of other chemical mutagens (Samoiloff and Smith 1971; Pertel 1973; Cadet and Dion 1973; Person 1974; Person and Brun 1974; Herman and Horvitz 1980). Spontaneous mutations have also been observed in *C. elegans*, and two dwarf mutations at different loci in *C. elegans* Bergerac were studied by Dion and Brun (1971).

The most easily observed mutations are those affecting size and shape, locomotion, or behavioural responses to stimuli. These have been described by such terms as dumpy, blistered, long, roller, uncoordinated. Brenner (1974) identified 300 visible mutants distributed on all six linkage groups. The linkage groups corresponding to the five autosomal chromosomes designated by Latin numerals I to V, and the sex chromosome, X, making up the haploid complement. The males are 5AA + XO and females are 5AA + XX. The following terminology in general use is taken from Herman and Horvitz (1980). Mutations are given a name with a one- or two-letter prefix in italics, designating the laboratory of origin, and a number, e.g. *el* 20. Genes (or complementation groups) are named by a three-letter italicized prefix, descriptive of the mutant allele, followed by hyphen and italicized number, e.g. *unc-4* (for uncoordinated). A mutant, when located in a complementation group can be described thus: *unc-4* (*el* 120)II, where II is the linkage group. The phenotype would be Unc-4 and the normal (wild type) non Unc.

Genes are defined in practice by complementation groups, based on complementation tests of closely linked recessive mutations giving similar phenotypes. Experimental crosses give progeny carrying both mutations, but on different homologous chromosomes. When these restore the normal or wild-type phenotype these mutations are said to complement one another and to affect different genes. When they fail to complement one another they are considered different alleles of the same gene. Difficulties arise with intra-cistronic mutations, and with sex-linked mutations. Because many of the mutations which give fertile homozygous hermaphrodites, render homozygous males infertile, heterozygous males must be used for experimental crosses. This is especially true of dwarf mutations, but ways have been found of mapping the position of closely spaced dwarf mutations in *C. elegans* Bergerac (Ouazana and Brun 1975). Brenner (1974) used temperature-sensitive mutations for mapping genes and complementation tests.

Establishing linkage groups and mapping the position of genes on chromosomes follows conventional genetic methods, with modifications because self-fertilization and cross-fertilization can both occur in the hermaphrodite. Easily identifiable recessive genetic markers are required, distributed along each of the chromosomes, and such mutants are available in both races of *C. elegans* (Brenner 1974; Ouazana 1974). With a new mutation its linkage group is established by examining the progeny of hermaphrodites heterozygous for both the mutant and markers (a + / + b). The double mutant (ab / ab) will occur with a frequency of 1/16 when these genes are on different chromosomes. A lower frequency indicates linkage, the double recessive arising from recombination. Recombination frequencies form the basis for mapping genes on chromosomes. The procedures to be followed in establishing linkage groups, mapping linkage groups, and recognizing deficiencies, duplications, and non-disjunction mutants in *C. elegans* are given by Herman and Horvitz (1980), who also give a map locating a large number of genes on all six chromosomes of *C. elegans*.

Mutations affecting the musculature of *C. elegans* have proved particularly valuable for studying the molecular biology of contractile proteins (see Section 2.6) and many of these are described by Zengel and Epstein (1980). Others affecting behaviour have been used to investigate sensory physiology of *C. elegans* (Section 3.5). Temperature-sensitive mutants have proved of great interest in developmental studies. In such mutants growth and reproduction are normal at the *permissive* temperature and abnormal at the *restrictive* temperature. Hirsch and Vanderslice (1976) have studied temperature-sensitive mutants t_s induced by EMS, with 16 °C as the permissive temperature and 25 °C as the restrictive temperature. Cold-restrictive mutants are also known. By varying the stage of development at which the nematode is exposed to a change in temperature, different metabolic and growth disturbances can be defined. The authors found 223 different t_s mutants affecting every stage of growth and maturation.

Three autosomal t_s mutants blocking embryogenesis at the restrictive temperature (25 °C) were selected for more detailed analysis (Vanderslice and Hirsch 1976), i.e. zyg-1, zyg-2, and zyg-3. By crossing males carrying t_s zyg-1 with hermaphrodites carrying a dumpy marker, zyg-1 was located on chromosome II. Eggs from heterozygous parents overcome the developmental blockage, but homozygous t_s eggs fail to develop at the restrictive temperature, illustrating maternal effects.

The action of 25 t_s zyg mutants during development was further investigated by Hirsch (1979). With 21 of these, homozygous progeny segregated from self-fertilizing hermaphrodite parents, which developed at the restrictive temperature, so that expression of the normal allele in the parent was sufficient, i.e. maternal effect was sufficient for development. In another test, homozygous parents, raised at 16° to a late developmental stage and then raised to 25° were crossed with normal males. With 13 out of 25 mutants, the gene or cytoplasm from the spermatozoa did not compensate for the deficiency in the parent of the maternal effect gene.

With three mutants, expression of the normal gene was required in the zygote, while with one mutant, both maternal and zygote expression was necessary.

The Bergerac and Bristol races of *C. elegans* differ in their upper temperature limits, though it is possible to raise the maximum temperature for the more sensitive Bergerac race to that of the Bristol race, provided this is done gradually over many hundreds of generations (Brun 1965, 1966*a*, *b*, *c*; 1967). Temperature sensitivity segregates in crosses between the races as a mendelian recessive (Fatt 1964; Fatt and Dougherty 1963). Within the Bergerac race genetic analysis of crosses between adapted and unadapted lines point to a dominant autosomal gene(s) conferring tolerance (Brun 1972), though sensitivity is influenced by other cultural factors (Fatt 1967; Brun 1967). Lethal and sterilizing t_s mutants, produced by EMS, have also been described in the Bergerac race of *C. elegans* (Abdulkader and Brun, 1978, 1980).

Mutagenesis by chemicals and X-irradiation has been studied in the fungal-feeding aphelenchid *Aphelenchoides composticola* (Person 1974; Person and Brun 1974). Dwarf and giant mutants were induced by EMS and X-irradiation, but some chemical mutagens gave rise only to giants.

8. Ecology

8.1. Introduction

NEMATODES are essentially aquatic, dependent on at least a film of moisture for activity, though many inhabit periodically arid and frozen lands, surviving desiccation or freezing in a quiescent or cryptobiotic state. Because most live within the interstitial spaces between the soil particles, or other substratum, their ecology is dominated by such physical properties as porosity, viscosity, surface tension, gaseous diffusion, water percolation, and humidity. I shall begin by discussing the ecology of terrestrial nematodes separately from marine nematodes because of the differences in their environments and because of the different methods which must be used to study them. No sharp distinction can be drawn between the ecology of terrestrial and freshwater nematodes, though I have thought it worthwhile devoting a short section to nematodes from lakes.

Having discussed the ecology of these three groups separately, I shall try to find as much common ground as possible. Bacteria-feeding nematodes are important in all nematode environments and a section will be devoted to bacteria-feeding. The contributions made by nematodes to the flow of energy through terrestrial, freshwater, and marine environments will be compared. The importance of nematodes in agriculture and forestry has stimulated a great deal of work on the microbial pathogens and predators of nematodes. Nematode-trapping fungi in particular have been the subject of many fascinating papers. Nematodes are themselves disseminators of microbial pathogens of animals and plants. The relationship between soil nematodes and other organisms is discussed, but I can find relatively little on these subjects in aquatic nematodes. Agricultural nematology is not discussed, but the reader will find nematode ecology, with special reference to agriculture, the subject of books by Norton (1978) and Wallace (1973). No attempt will be made to summarize faunal surveys which would require a book in itself, but references are given to faunistic work from a wide variety of habitats excluding agricultural land.

Nematologists face much greater practical difficulties in analysing marine nematode communities than terrestrial communities. One of the less obvious difficulties is that few nematodes from offshore communities have been cultured in the laboratory, so that many aspects of their behaviour cannot be directly observed. Marine nematologists have made much use of diversity indices and experimented with mathematical models of population structure as less direct means of studying the relationships

163

TABLE 8.1

Abundance and biomass of nematode fauna from different types of ecosystems

Ecosystem	Abundance			Biomass (mg fw m^{-2})			
	No. of sites	Thousands m^{-2}		Sites	Mean value	Range	ind. (μg fw)
		Mean value	Range				
Tundra	15	3490	800–10 000	11	1350	265–4130	0.38
Coniferous forest	14	3330	1125–15 000	9	510	180–1696	0.21
Eucalyptus forest	4	5467	4040–7449	4	1423	770–2050	0.26
Deciduous forest	15	6270	255–29 800	9	2760	75–15 200	0.27
Temp. grassland	20	9190	2432–30 000	12	3800	650–17 800	0.52
Fens, bogs, heathland	9	1660	330–3900	4	660	350–900	0.31
Deserts	2	760	423–1100	2	410	125–700	0.47
Tropical forests	1	1700	1500–1900	—	—	—	—

Reproduced from Sohlenius (1980).

between communities and their environments. Nematode ecology presents challenging problems, not the least of which is great diversity of species and large numbers of individuals which are found in small samples of soil, sand, or mud. Often closely related species with similar habits and food occur together, raising questions about competition, resource partition, and succession.

8.2. The distribution and abundance of nematodes in the soil

All kinds of soils in every part of the world contain large taxonomically diverse populations of nematodes. The nematodes are usually most abundant near the surface, with the majority within the top 10 cm, though some may be found much deeper. For example, a small proportion have been found below 30 cm in grassland (Yeates 1980), between 20 and 30 cm in temperate forests (Volz 1951), and between 30 and 40 cm in coastal sand dunes (Yeates 1968). Populations are densest in zones rich in organic matter, or where fine plant roots are concentrated. Adequate soil moisture and oxygen favour dense populations.

Two recent authoritative reviews have discussed the abundance, biomass, and distribution of soil nematodes (Yeates 1979b; Sohlenius 1980) and I shall rely heavily on these reviews in this section. For such work the population density of each species throughout the year must be estimated from soil samples. Estimates of biomass can be made by summing the numbers and volumes of each species over the year, and their contribution to total soil respiration can be calculated from laboratory measurements. Despite the formidable task, many such surveys have now been reported, though it is difficult to obtain publications. I have reproduced summaries published by Sohlenius (1980) in Tables 8.1 and 8.2. Others have been published by Yeates (1979b). Examples from a variety of habitats which can be found in widely distributed scientific journals are reproduced below:

Temperate forests and woodland	Volz (1951); Bassus (1962); Yeun (1966); Wasilewska (1970); Yeates (1972b); Sohlenius (1979)
Grasslands	Yuen (1966); Kimpinski and Welch (1971); Yeates (1974, 1978a, b, 1980)
Moorland	Banage (1963)
Tundra	Chernov, Striganova, and Anajeva (1977)
Antarctic Island	Caldwell (1981)
Tropical swamps	Banage (1964)

TABLE 8.2

Number, biomass, and respiratory metabolism of nematode fauna from different localities. Figures show m^2 values for abundance biomass and energy liberation by respiration $(kCal\ y^{-1})$

Site	Abundance (thousands m^{-2})	Biomass (g fw m^{-2})	Respiration (kCal m^{-2})
Philipson *et al.* (1977)			
Beech forest England	370	0.08	1.4
Freckman and Mankau (1977)			
Nevada desert	420	0.13	7.6
Yeates (1972, 1977)			
Beech forest Denmark	1430	0.37	5.2
Wasilewska (1971)			
Pine forest 10-y old Poland	1580	0.31	17.1[1]
Wasilewska (1974*a*)			
Pasture Poland	3500	2.2	50.0
Lagerlöf *et al.* (1975)			
Tundra Sweden	4090	1.09	7.3
Sohlenius (1979)			
Pine forest Sweden	4110	0.58	8.2
Nielsen (1961)			
Grassland St. 2 Denmark	5000	10.0	115.0
Wasilewska (1971)			
Mixed forest 17–20-y old Poland	6950	0.69	33.0[1]
Wasilewska (1974*b*)			
Rye field Poland	8620	1.06	60.0[1]
Nielsen (1961)			
Grassland St. 18 Denmark	10 000	14.0	339.0

[1] Recalculated values (Wasilewska personal communication).
Reproduced from Sohlenius (1980). (Please see original journal for details of references listed here.)

Comparisons are difficult because of the differences in the efficiency of the methods used. With the Baemann-funnel technique, for example, eggs may hatch or cryptobiotic stages become active, especially with surveys from deserts and subpolar regions, which would not be recovered by sieving, complicating comparisons. I have not included surveys from argicultural land under crops, but with the exception of the tundra survey, all those cited are from lands under management or subject to some human interference. Yeates (1977*a*, 1978*a*, 1979*a*, *b*) discusses effects of pasture management in some detail.

In one of the most thorough of the examples, Yeates (1972*b*) sampled a Danish beech forest at monthly intervals throughout the year, finding 75 species, with a maximum population in May of 1.45×10^6 m^{-2} and a minimum in February of 0.40×10^6 m^{-2} (annual mean 1.09×10^6 m^{-2}). Low temperatures in winter probably restricted recruitment, while dryness in summer may have reduced numbers. Biomass was also a minimum in February (0.93 g m^{-2}), but a maximum in November, 5.10 g m^{-2} (mean annual = 0.28 g m^{-2}). In another example, in a Swedish coniferous forest Sohlenius (1979) found a winter maximum of 6.3×10^6 m^{-2}, with a mean annual abundance of 4.11×10^6 m^{-2}. The numbers in winter may have been influenced by emergence of active nematodes in the laboratory in soil samples.

Populations in temperate grasslands and deciduous forests are high; populations are lower in coniferous forests and lower still, to judge from the few examples available, in tropical forests, but the tropical forests may be less influenced by seasonal variations. Even in the polar desert of Northern Siberia (Taimyr Peninsula) numbers and biomass are high, 1–8 g m^{-2}, from 100–40 000 ind. m^{-2}, with population density dependent on the plant cover of mosses and lichens (Chernov et al. 1977). A total of 53 species were found. The mean annual populations may be ecologically less significant than the seasonal maxima, and Yeates (1979b) gives the following ratios between seasonal minima and maxima: English beech woodland 2×; New Zealand sand dune 3×; French grassland 4×; English moorland and Danish beech forests 5×; Finnish and Swedish pine forests 6×; New Zealand pasture 10×.

Nematodes show very clearly the distinction some ecologists have drawn between r-strategists and k-strategists (Johnson, Ferris, and Ferris 1974). The r-strategists have short generation times and high fecundity, rapidly exploiting transient resources. Many bacteria-feeding nematodes fulfil this roll and their numbers fluctuate widely in time and space, so that the numbers in surveys may be influenced by transient circumstances, e.g. a fortuitous dung pat from a grazing animal or the death of a plant. *C. elegans*, a hermaphrodite with a generation time as short as 2½ days and each individual capable of producing 280 eggs, is an r-strategist. When food is scarce, it gives rise to persistent dauer larvae. Other nematodes are k-strategists, with long generation times and with perhaps a single generation a year, and low fecundity. Some Dorylaimida and Tylenchida fulfil this roll. Examples of different generation times are given in Table 8.3.

Many surveys have classified the nematode fauna into feeding categories, and I have reproduced some examples in Table 8.4. However, such classifications are probably too imprecise to be very useful. Tylenchida can with some confidence be classified as plant feeders or fungal feeders on the basis of their stomatostyle, but this order also includes insect parasites and predators as well (Siddiqi 1980). Even within the plant feeders, it is ecologically important to distinguish between endo- and ecto-parasites

TABLE 8.3

Examples of generation times and fecundity in free-living nematodes under favourable conditions

Taxon and species	Generation time egg–egg	Eggs/ female	Temp. C	Normal habitat	Diet	Authors
RHABDITIDA						
Caenorhabditis elegans	96 hrs	280	19.5	soil	bacteria	Byerly et al. (1976)
Rhabditis marina	5 days	70–100	20	marine, littoral	bacteria	Tietjen et al. (1970)
Mesodiplogaster lheritieri	2–3 days	570	25	soil	bacteria	Grootaert (1976)
Mononchoides potohikus	9 days	146	20	sand dunes	bacteria	Yeates (1970)
Paroigolaimella bernensis	46–8 hrs	90–100	30	sewage	bacteria	Pillai and Taylor (1968)
TYLENCHIDA						
Aglenchus costatus	27–35 days	62–105	18–20	soil	plant roots	Wood (1973c)
APHELENCHIDA						
Aphelenchus avenae	6 days	—	28	soil	fungi	Hechler (1962b)
Bursaphelenchus fungivorus	5–6 days	121–214	20–5	soil	fungi	Franklin and Hooper (1962)
MONHYSTERIDA						
Monystera disjuncta	10–12 days	—	25	marine mud	bacteria	Tietjen and Lee (1972)
Diplolaimelloides sp.	7 days	—	24	marine mud	bacteria	Hopper et al. (1973)
CHROMADORIDA						
Chromadora macrolaimoides	22 days	9–11	25	marine algae	algae	Tietjen and Lee (1973)

ENOPLIDA

Enoplus communis	1 year	10	—	marine, littoral	omnivor	Wieser and Kanwisher (1960)
Oncholaimus oxyuris	101 days	36	25	marine, littoral	omnivor	Hiep *et al.* (1978)
Mononchus aquaticus	14–15 days	—	28	freshwater	predator	Grootaert & Maertens (1976)
Aporcelaimellus sp.	95–130 days	—	18	soil	algae	Wood (1937*a*)

TABLE 8.4

Some examples of the relative abundance (per cent) of different feeding categories of nematodes in soils

| Food category | Moorland,[1] Britain | | | | Britain[2] | | Sweden[3] |
	Calluna	Juncus	Nardus	Grass	Woodland	Grassland	Pine forest
Plant root/fungal	82	65	60	48	36	57	36.5
Microbial/bacterial	12	29	25	25	46	18	56.5
Omnivorous	—	—	—	—	11	17	—
Predatory	0	0	0.2	1	1.1	1	—
Miscellaneous/other	1	4	8	21	—	8	7

[1] Banage (1963).   [2] Yuen (1966).   [3] Sohlenius (1980).

whose distribution is closely tied to appropriate host plants and those that move through the soil browsing on a variety of plant roots. Within the Dorylaimida the difficulties are greater because feeding habits are wider and less well known. Closely related morphologically similar taxa include algal feeders, plant-root feeders, and predators on nematodes. The Mononchidae are usually recognized as predators, feeding voraciously on other nematodes, but the same species also feed on protozoa and aggregations of bacteria in the laboratory, and probably require bacteria as well as animal food in their diet. Yeates (1971) has suggested classifications, but the feeding habits of many common species are insufficiently known.

Another way of analysing soil nematodes is to look for similarities and dissimilarities in the taxonomic spectrum and associate these with differences in other ecological factors. Johnson, Ferris, and Ferris (1972) used a computer program to analyse the similarities in species found at 18 isolated patches of woodland on agricultural land in Indiana, USA. Altogether 175 species were found in the soil sample, with many of the species widely distributed, 18 occurring in all the woods and nearly half the species in 50 per cent of the woods. The program, which expressed similarities in the form of a dendrogram, showed that the terminal branches separated similarities associated with different vegetation and soil types. In further work (Johnson, Ferris, and Ferris 1973, 1974), extending over two years, community ordination statistics were used, which also took into account the numbers of each species present in samples. The 'prominence' of each species was calculated, a statistic adopted from plant ecology: Prominence $= p\sqrt{n}$, where p is the total of individuals of a species at a given site and n is the total number of all species at site. It has also been used by Knobloch and Bird (1978) in surveys of plant-feeding species, especially Criconematidae.

Johnson *et al.* (1973, 1974) showed by community ordination statistics that it was possible to associate the distribution of particular nematode species with the interrelated combinations of the type of plant cover and soil. Other similar surveys restricted to plant parasites could be quoted which demonstrate associations between nematode distributions, vegetation, and soil types, but these do not tell us much about which factors influence the distributions.

Some comparisons can be made between similar habitats in different parts of the world; beech forests in Denmark (Yeates 1972*b*), Germany (Volz 1951), and beech woods in Indiana, USA (Johnson *et al.* 1972); pine forests in Sweden (Sohlenius 1979) and Poland (Wasilewska 1970); grasslands in New Zealand (1978*a, b*), Canada (Kimpinski and Welch 1971), and Denmark (Overgaard-Nielsen 1949). Some genera and some species are very widely distributed throughout the world and with different vegetation. This is true of bacteria-feeding nematodes associated with decaying plant remains, i.e. Cephalobidae *Eucephalobus, Acrobeles, Acrobeloides, Cervidellus, Cephalobus*; Tetracephalidae *Tetracepholus*; Panagrolaimidae *Panagrolaimus*; Plectidae *Plectus* and *Wilsonema*. Several genera associated with the moss, liverwort, layer: *Aphelenchoides, Monhystera, Tylenchus, Eudorylaimus*, and *Prismatolaimus* are also very common. The bryophyte fauna becomes the total fauna in the mosses on Antarctic Islands, with several cosmopolitan genera (Caldwell 1981). The Rhabditida are more taxonomically diverse, but common, and are probably bacteria feeders associated with faecal material, dead invertebrates, and other decomposing material. The Dorylaimida with varied food habits and plant-feeding Tylenchida are usually represented by more numerous genera, but are taxonomically much less predictable. Several genera of Criconematidae are commonly associated with forests and woodlands. Hill moorlands in Britain (Banage 1963) show close affinities with marshes elsewhere, probably because the soils are saturated with water.

Many nematode genera and some species have world-wide distributions, but whether the major zoogeographic regions have distinctive fauna is not clear. Many of those associated with agricultural land have been widely distributed by man (Norton 1978) and some are transported by wind (Carrol and Viglierchio 1981). Ferris, Goseco, and Ferris (1976) suggest that the distribution of Leptonchoidea (Dorylaimida) shows evidence of separate evolutionary radiation after the break-up of Gondwanaland in the Mesozoic. This is the only paper I have come across relating nematode faunas to plate tectonics. Sudhaus (1974*a*) found different geographical races of *Rhabditis marina* on European shores (marine, but from a typically terrestrial family). *R. oxycerca* from Australia interbreeds with *R. oxycerca* from Europe, and has a world-wide distribution (Sudhaus 1980*b*). The Antarctic has species known only from that continent, but from cosmopolitan genera (Yeates 1979*c*). Yeates (1973*b*) found six new species amongst 30 collected from Vanuatu Islands, but no new genera. Of those species known from elsewhere, 11 are cosmopolitan in their distribution, seven are tropical, and the remainder are from elsewhere in the southern hemisphere or with a poorly known distribution.

8.3. The physical properties of the soil and their effect on nematodes

The dimensions of the soil spaces between the soil particles directly and indirectly affect nematodes, which must move in films of moisture over the soil particles and pass through the narrow channels between them. These relationships have been studied by Wallace whose work on nematode locomotion has already been referred to (Section 3.2) and I shall base my discussion of the physical properties of the soil largely on his account (Wallace 1971). The size of the particles and their packing determines the channels open to nematode movement, which is most efficient with particles about ⅓ the diameter of the nematode length (Wallace 1958c). They can be too narrow, as in compact clays, excluding nematodes. The dispersal of larval *Heterodera schachtii*, for example, which are 15–18 μm in diameter, is restricted when the channels have equivalent pore diameters of less than 30 μm (defined below), and stopped when less than 12 μm (Wallace 1958a). The channels may also be too wide, because there may not then be a continuous film of water (Wallace 1958b).

The channels between the particles can be described in terms of equivalent pore diameter, assuming the 'necks' between the particles to be

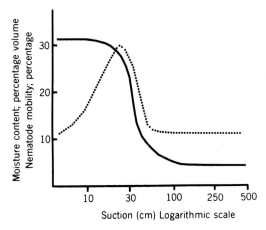

FIG. 8.1.   The moisture characteristic of a soil sample and the associated mobility of terrestrial nematodes in the soil. Nematode mobility is the relative rate of dispersal of nematodes over a given distance when added to the soil. The hypothetical graph is derived from experimental results published by Wallace (1958b).

circular. A frequency distribution of equivalent pore diameters for a soil sample can be calculated from the moisture characteristic (Wallace 1971). The moisture characteristic expresses in graphical form the progressive draining of water from a saturated sample as it is exposed to increasing suction pressure (see Fig. 8.1). Methods for making these measurements can be found in textbooks of soil science (Wallace 1958a). Assuming the 'neck' to be circular, the suction can be related to the surface tension at the necks

$T\pi d = S\pi(d^2/4)$ or $d(\mu m) = 3000/S(cm)$,
where T is the surface tension at a soil/air interface, S is the suction pressure, and d is the equivalent pore diameter.

In soil with uniformly sized particles, about ⅓ of the volume can be occupied by either air or water. In soil containing unsorted particles of widely different sizes, the volume will be decreased by small particles filling in between larger ones. Organic crumbs, unlike mineral particles, shrink or swell according to the degree of hydration. The degree of saturation does not only affect nematodes by providing films of water in which to move, but also by exerting forces on the nematode's body and by determining the availability of oxygen. The rate of diffusion of oxygen is much greater in air than water, so that in saturated soils microbial activity renders soils anaerobic. Soil nematodes become quiescent in anaerobic conditions (see Sections 3.6 and 3.8).

The moisture characteristic, used by Wallace to measure the channels open to nematode movement, also provides a measure of the capacity of a soil to hold water. Water is lost by evaporation and drains by gravity. It is retained by capillary action (i.e. surface tension), absorption by organic matter and on clay particles, and to a lesser extent by osmotic pressure (except in saline soils). The moisture profile is strongly influenced by pore sizes, and generally shows decreasing water content with increasing height above the water table. The profile is an important factor in the distribution of nematodes with depth, and the movement of water through the soil aids the dispersal of nematodes. The matric potential of a soil is the sum of surface tension and absorptive forces tending to retain water, and is measured by the suction pressure in equilibrium with forces retaining water. The sum of the matric potential and the osmotic pressure gives a measure of the free energy ψ of water in the soil, usually expressed on a logarithmic scale

$$pF = \log \{-\psi(cm)\}.$$

Plant growth slows at pF 3.3 and plants wilt at pF 4.0. The dispersal of the larvae of *Ditylenchus dipsaci* is restricted at pF 3.87 and stops at pF 4.35 (Blake 1961). Nematodes from the Mojave Desert responded to increasing suction pressure by entering cryptobiosis at well below the wilting point of plants (Demeure, Freckman, and Van Gundy 1979a). Survival, however, which requires internal physiological responses is critically dependent on the rate of dehydration (Demeure et al. 1979b).

8.4. Nematodes and fungi. Predaceous fungi, microbial pathogens and predators of nematodes

Nematodes which will feed on fungal mycelia in laboratory cultures are common in the soil and have also been found in coastal marine habitats. I have discussed the culture of fungal-feeding Tylenchida and Aphelenchida in Section 4.2, noting that some, like *Ditylenchus destructor*, have a very wide taxonomic range of potential food fungi, and may even feed on higher plant tissues as well. Common fungal-feeding species belong to the following genera: *Aphelenchoides, Aphelenchus, Paraphalenchus, Bursaphelenchus, Neotylenchus, Deladenus, Hexatylus*, and *Ditylenchus*. Some feed on mycorrhizal fungi in the laboratory and are associated with the mycorrhizal fungi on plant roots, though it is not clear whether they harm the plant (Clark 1964; Riffle 1968; Norton 1978). Several species are important pests of commercial mushroom beds, while others are associated with bark beetles (Massey 1974) and Hymenoptera (Bedding 1968) and transported by insects to feed on the associated fungi. Nematodes from other orders feed on fungal spores, e.g. Dorylaimida (Hollis 1957), Rhabditida (Hunt and Poinar 1971; Gupta, Singh, and Sitaramaiah 1979; Barron 1977) and Aphelenchida (Gupta et al. 1979). The spores of plant pathogenic fungi and bacteria are disseminated by bacteria-feeding nematodes by passing through the alimentary canal unharmed (Jensen 1967; Jensen and Siemer 1969).

However, I want now to turn to the fungi which attack nematodes. The specialized traps developed by some fungi to entrap nematodes have been the subject of many publications, but nematode-trapping fungi are not the only ones to attack nematodes. An excellent and well illustrated booklet by Barron (1977) describes the various kinds of fungi attacking nematodes and their biology. This booklet will introduce the reader to the whole subject with a full bibliography. Other general reviews have been published by Dreschsler (1941), Duddington (1955), Cooke (1968), and Mankau (1980), all of whom have contributed to the subject. A taxonomic key has been published by Cooke and Godfrey (1964).

It is useful to distinguish between predatory fungi, that produce an extensive hyphal system which invades entrapped nematodes, and endoparasites, in which the hyphae are confined to the nematode's body and short reproductive hyphae. The endoparasitic fungi produce infective spores, which persist in the soil, or as in some phycomycetes, infective motile flagellated or amoeboid zoospores. The two categories are not entirely clear-cut, with some genera having species falling into both categories. The richest source of both kinds appears to be farmyard soils (Barron 1977) with agricultural land (Duddington 1951, 1954; Wood 1973b) and forest soils (Capstick, Twinn, and Waid 1957) being good sources. The common feature is that nematode-destroying fungi are most numerous where there are concentrations of decaying organic matter. The only references I have found to aquatic species are to four species of *Dactylaria* from ponds in Britain (Peach 1950, 1952).

Endoparasitic fungi have been reported from the classes: Chytridiomycetes, Oomycetes, Zygomycetes, Deuteromycetes, and Basidiomycetes (Barron 1977). One of the commonest and best known is *Catenaria anguillulae* (Chytridiomycetes), which produces motile zoospores that encyst on nematodes, which are then invaded by its hyphae. *Catenaria* attacks rotifers and tardigrades as well as many kinds of nematodes and is common in a wide range of decomposing organic matter. Other endoparasites invade the nematode body from spores which adhere to the nematode's body on contact. *Haptoglossa heterospora* liberates non-motile spores which can inject an infective particle through the cuticle of a nematode by a minute tube (Barron 1977). Endoparasitic fungi with spores that adhere to the cuticle of nematodes generally have very tiny spores, about 2 μm in diameter, while others, which have spores which must be ingested by the nematode before they can invade their hosts, usually have larger spores of complex shape, such as crescents or helices, which lodge in buccal cavity or oesophagus. Aphelenchida, Tylenchida, and Dorylaimida are probably not susceptible to oral infection because of their narrow feeding stylets. Nematode-trapping predaceous fungi may produce traps spontaneously along their extensive hyphae, or only in the presence of nematodes. In a few the whole mycelium is adhesive (Zygomycetes), but more usually specialized adhesive structures such as erect branches, knobs, or two- or three-dimensional networks are formed. Nematodes may also be trapped in non-constricting or constricting rings. The constricting rings are formed from three cells, which rapidly close on a nematode which enters the ring. Constricting rings can be stimulated to contract artificially by warmth or contact. A rapid uptake of fluid by the cells is involved in ring closure, but the mechanism is controversial. Most predaceous fungi are Deuteromycetes (though one Basidiomycete with adhesive knobs is known).

Little is known about the nature of the adhesive, though in *Arthrobotrys oligospora* experiments have demonstrated a 'lectin' on the adhesive network which binds selectively to carbohydrates on the nematode cuticle probably initiating changes in the cuticle (Nordbring-Hertz and Mattianson 1979; Jansson and Nordbring-Hertz 1980).

Predaceous fungi may release toxins when nematodes are entrapped (Olthof and Estey 1963), nematode attractants (Jansson and Nordbring-Hertz 1980), and antibiotics. Balan and Gerber (1972) found *Arthrobotrys dactyloides* liberated sufficient ammonia into culture media to kill *Panagrellus redivivus* as well as releasing attractants. Barron (1977) showed that two endoparasites, *Harposporium anguillulae* and *Meria coniospora*, released antibiotics which suppressed the growth of several common soil saprophytic fungi. Trap formation by some species in culture may require the presence of nematodes or an alternative stimulating agent. Pramer and Stoll (1959) found cultures of *Neoaplectana glaseri* produced a stimulant described as nemin which induced trap formation in *Arthrobotrys conoides*. Amino acids or peptides are constituents of nemin, but induction involves more than nemin (Nordbring-Hertz 1977), and nematode movement is also a stimulant (Jansson and Nordbring-Hertz 1980).

The advantage gained by nematode-trapping fungi over other saprophytic fungi is not clear, because, unlike the endoparasitic species, they can grow in the absence of their prey. Cooke (1963, 1968) believes that they compete poorly with other saprophytes in the absence of prey, though which nutrients are important is unknown. Perhaps vitamins are made available. Anyway, Cooke concludes, from studies of their prevalence in soils, that nematode-trapping fungi only become common at specific stages in the decomposition of organic matter, and that their hyphae and spores do not persist for more than a few weeks. Soil amoebae feed on fungal spores (Barron 1977). This probably refers to infective conidia, because both predaceous fungi and endoparasites produce dormant resting spores as well as infective spores in culture and presumably in the soil. The nematophagus fungi may not always have the advantage. In culture *Aphelenchus avenae* fed on five predaceous species, and though some nematode-trapping occurred, the nematode destroyed the fungi without decreasing in numbers itself (Cooke and Pramer 1968). Little definite can be said about specificity, but most nematode-destroying fungi appear to have a wide range of potential prey.

A number of attempts have been made to utilize fungi for the biological control of plant nematodes, generally with little success (Cooke 1968; Mankau 1980), though Mankau reports that Cayrol and others have successfully used *Arthrobotrys*, a predaceous genus, to control *Ditylenchus myceliophagus* in commercial mushroom beds and against *Meloidogyne* on tomatoes (cited by Mankau 1980). However, fungi may play a very significant role in limiting populations of cyst nematodes, *Heterodera*, in agricultural land without human intervention (Tribe 1977, 1979; Kerry 1980). Fungi attack the female on the host plant, for example endoparasitic *Catenaria auxiliaris*, and the eggs in the cyst, for example *Verticillium chlamydosporium* and 'black yeast'. The eggs of free-living nematodes are also attacked by fungi, e.g. *Rhopalomyces elegans*, a predaceous form.

Though fungi are the most prevalent microbial destroyers of terrestrial nematodes, other parasites have been described. The best known is now generally referred to as *Bacillus penetrans*, though it may be an actinomycete (unusual in forming endospores), and was until recently generally considered a protozoan under the name of *Dubosqia penetrans* (Sayre 1980; Mankau and Prasad 1977). Its taxonomic position remains doubtful (Sayre 1980; Mankau and Imbriani 1975; Imbriani and Mankau 1977), but it is certainly a prokaryote. It is a widespread pathogen of plant-feeding nematodes in North America, which it infects through the cuticle by adhesive spores which give rise to a septate hyphal mycelium invading the pseudocoelom (Mankau and Imbriani 1975). The hyphae are like those of actinomycetes (Sayre 1980), but the spore resembles that of a bacterium (Mankau 1977; Mankau, Imbriani, and Bell 1976). Not all the nematodes tested as hosts proved susceptible (Mankau and Prasad 1977).

A number of other bacterial, rickettsial (Banage 1965; Shepherd, Clark, and Kempton 1973; Sayre 1980), and protozoal (Canning 1962, 1973) pathogens have also been reported but studied in less detail. *Anguina agrostis* (Tylenchida), which causes galls on rye grass in Australia, passively carries the bacterium *Corynebacterium rahay*, preventing the reproduction of the nematode and producing a toxin capable of killing sheep (Stynes 1980). Marine *Metoncholaimus scissus* is infected by a microsporidian (Hopper, Meyers, and Cefalu 1970).

The transmission of viruses is a feature of two taxa within the Dorylaimida. *Longidorus* and *Xiphinema* (Dorylaimoidea) transmit a number of viruses of the NEPO group to plants, while *Trichorodorus* (Diphtherophoroidea) transmits TOBA viruses to plants. Virus transmission by these nematodes has been the subject of a symposium published as a textbook (Lamberti, Taylor, and Seinhorst 1975). Interestingly, successful transmission depends on the retention of virus particles, ingested by the nematode when feeding on an infected plant, at a specific region of the alimentary canal, which differs in the three genera (Harrison, Robertson, and Taylor 1974). In *Longidorus* the isometric RNA NEPO viruses are held in the odontostyle region, while in *Xiphinema* they are retained in the oesophagus and odontophore lumen. In *Trichodorus* the rod-shaped RNA TOBA viruses are retained in the odontostyle and oesophagus. The virus particles can be found in these regions by electron microscopy. There are differences between the species of each of these genera in their capacity to transmit particular virus diseases.

Virus-like particles have been observed in the tissues of *Dolichodorus heterocephalus* (Zuckerman *et al.* 1973*b*). The bacteria-feeding nematode *Mesodiplogaster* (=*Pristionchus*) *lheritieri* can transmit bacterophage to

Streptomyces griseus (Jensen and Gilmour 1958) and to *Agrobacterium tumefasciens* (Chantanao and Jensen 1969) in the laboratory. The same nematode transmits vertebrate-cell mycoplasma in laboratory cultures (Jensen and Stevens 1969). *M. lheritieri* feeds on plant-pathogenic bacteria (*Agrobacterium tumefasciens, Erwinia amylovora, Pseudomonas phaseolicola, E. carotovora*), soya bean rhizobial bacteria (*Rhizobium japonicum*), and the spores of plant-pathogenic fungi (*Fusarium oxysporum, Verticillium dahliae*), passing some viable organisms through the alimentary canal so that these can be disseminated to susceptible plants (Jensen 1967; Jensen and Siemer 1969; Chantanao and Jensen 1969; Jatala *et al*. 1974).

Many nematodes are predatory on other nematodes and common in the soil. Several examples which can be cultured in the laboratory have already been discussed in Chapter 4. The Mononchidae, a large family, are very common. Other predatory genera belong to the Dorylaimidae, Aporcellaimidae, Ironidae, Aphelenchoididae, and Diplogasteridae. Teeth, stylets, and jaws are used to puncture the prey, which in the Mononchidae are swallowed whole. The Nygolaimidae feed on oligochaetes. Nematodes are also preyed upon by tardigrades (Sayre 1969), mites (Rodriguez, Wade, and Wells 1962), turbellaria (Sayre and Powers 1966) and amoebae (Sayre 1973). Circumstantial evidence suggests that predatory mites control the numbers of bacteria-feeding nematodes in some arid soils (Whitford, Freckman, Santos, Elkins, and Parker 1982).

8.5. The nematodes of fresh water and inland saline waters

No sharp distinction can be drawn between the nematode faunas of wet terrestrial environments and those of ponds and rivers and shallow lakes. Nematodes are essentially aquatic animals and the two faunas grade into one another. The fauna is densest where the organic content of the sediment is high, as in the presence of emergent plants, more sparse in sandy beaches. In bog soil and static freshwater the development of anaerobic conditions may limit nematode activity.

The benthos of deep lakes is interesting because of problems associated with thermal stratification and geographical isolation. The hypolimnion of lakes may become anaerobic for varying periods of the year. The nematode fauna is relatively limited, contrasting with the very diverse fauna of deep ocean sediments. Studies of Lake Tiberias in Israel, in which the hypolimnion is O_2-free for eight months of the year, by Por and Masry (1968), found large numbers of *Eudorylaimus andrassyi* at 43 metres, together with a tubificid oligochaete, a chironomid insect larva and a rhabdocoel platyhelminth, during the O_2-free season (winter). In Lake Champlain in North Eastern USA, the shallow water has a varied nematode fauna, but below 10 metres is largely restricted to one species

Ethmolaimus pratensis (Fisher 1968). Neither of these two benthic nematodes are restricted to deep lake benthos. *E. pratensis* has for example been reported from the Baltic (Jensen 1979*a*).

In Neusiedlersee, a large shallow lake in Austria, *Tobrilus gracilis* is associated with anaerobic zones several cm deep in the mud, beyond the limits of emergent vegetation. There is some indirect evidence (see Section 3.7) that this species is tolerant of anaerobiosis. Other nematodes occur near the surface of the mud, their total numbers fluctuating from 70 to over 600 m^{-2} in the mud, with a maximum in spring (Schiemer, Loffler, and Dollfuss 1969).

The nematode populations and their annual productivity have been studied in an alpine lake in Austria by Bretschko (1973). This lake is frozen over for the long winter, but nonetheless the water does not become anaerobic. The shallower reaches of the lake have a typical aquatic fauna, *Ironus tenuicaudatus*, *Tripyla glomerans*, *Tobrilus grandipapillatus*, *Mononchus* sp, and two *Dorylaimus* sp, which are probably predators and algal feeders. The deeper benthos, below 20 m, contains within the top 6 cm of silt *T. grandipapillatus*, *Monhystera stagnalis*, and *Ethmolaimus pratensis*. There were two to four generations per year in these species, and the total yield of nematode biomass was estimated at 66 kg per year or 4 kg per hectare. Production was higher in shallow water and during the period when the lake was ice covered (77 per cent between October and January; only 9 per cent from June to September). *I. tenuicaudatus* is abundant in shallow water—annual mean 13 300 m^{-2} (46.4 mg m^{-2}). *T. grandipapillatus* is also abundant, but in deeper water—235 000 m^{-2} under ice, 60 000 m^{-2} in summer.

The ecological isolation of lakes may be a factor in limiting the distribution of nematodes adapted to conditions in the benthos. The more diverse population in shallow water may be derived from terrestrial nematodes that can live in wet or marshy soils. The nematode fauna of Lake Baikal is of great interest because of its size, depth, and isolation. The shallow waters contain nematodes also found elsewhere, but the sediments below 200 metres have a unique nematode fauna (Tsaloliklin 1980). The genus *Tobrilus*, a common lake species, has given rise in Lake Baikal to many endemic species (a feature of other Baikal taxa). Tsaloliklin concludes that *Tobrilus* entered the lake when it formed (Miocene) from reservoirs, which had escaped the extinctions of other freshwater species. Later Dorylaimidae and Mononchidae invaded the lake, and later still other nematodes, e.g. Monhysteridae and Chromadoridae.

The nematodes of inland saline waters have been collected by a number of workers (Meyl 1955). Such waters contain marine brackish water and freshwater species according to their salinity. Thermal springs possess typical freshwater genera, but may contain locally adapted species, as in Yellowstone National Park, USA (see Section 3.8). A habitat which has attracted attention because of its extremely rich fauna of nematodes, though freshwater would be a misnomer, are various stages of sewage treatment plants (Peters 1930; Pillai and Taylor 1968; Murad 1970; Abrams and Mitchell 1978; reviewed by Schiemer 1975). Bacteria-feeding and predaceous Diplogasterinae are particularly prominent, with other families of bacteria-feeding Rhabditida well represented.

Some typically marine genera appear to have given rise to freshwater species as relict fauna of lakes. In Lake Nicaragua *Theristus setosus*, *Polygastrophora octobulba*, and *Viscosia papillata* var. *nicaraguensis* occur, of typical marine affinities (Meyl 1957). Riemann (1975) discusses a long list of nematode species from typically marine genera which have been reported from freshwater or terrestrial habitats. Species of *Axonolaimus*, *Adoncholaimus*, and *Enoploides* in the Don, Dneiper, Danube, and Volga appear to have entered the rivers from the Black or Caspian seas (Gagarin 1981). Marine nematodes may have become progressively adapted to freshwater by several routes; the progressive invasion of rivers from their estuaries, the invasion of coastal swamps, and isolation in arms of the sea which have become cut off and gradually diluted with freshwater, as with Lake Nicaragua. Perhaps less obvious has been migration from littoral marine sediments through the coastal ground water into subsoil waters. The fauna of 'ground' water associated with beaches has been discussed by Gerlach (1955). Superimposed is a fauna characteristic of supralittoral dunes (Gerlach 1953, 1967; Yeates 1967).

8.6. Marine nematodes

Nematodes are found in all kinds of marine sediments from tidal zones to the deepest ocean sediments. They are the most numerous and most diverse group of the meiofauna, often accounting for over 90 per cent of the individuals. The meiofauna can be arbitrarily defined as those animals small enough to pass through a sieve with a mesh of 0.5 mm diameter, but which are retained by a mesh of 50 μm. Nematodes are also associated with the algae on rocky shores and other marine plants, i.e. phytal nematodes. Although many genera of marine nematodes are very widely distributed there does seem to be some zoogeographical limitations according to (Tchesunova 1981). Species are more restricted in distribution than genera.

Most marine species have yet to be described, and the assignment of specimens to higher categories is often difficult. Consequently a crude classification into feeding categories has proved useful, making it possible to compare quantitatively, if not very precisely, the place of nematodes in the food web in different habitats. Wieser (1953, 1959*a*) assigned 361 genera to four feeding categories on the basis of observations on feeding habits and comparative studies of their mouth parts. It is possible to assign a specimen with some confidence to one of these categories from the structure of its head and buccal cavity. Wieser's categories, illustrated in Figs. 8.2 and 8.3 are:

1A. *Selective deposit feeders*, 97 genera. These have very small buccal cavities, or none, and feed by ingesting a fine suspension of particles, typically bacteria;

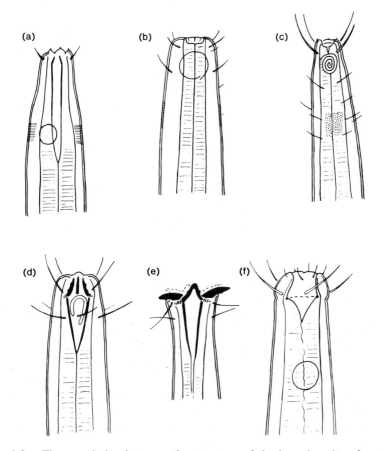

FIG. 8.2. The association between the structure of the buccal cavity of marine nematodes and feeding category. Group 1A selective deposit feeders: (a) *Cynura*; (b) *Terschellingia*; (c) aggregate feeder, *Sabatieria*. Group 1B non-selective deposit feeders; (d) *Odontophora* with teeth withdrawn; and (e) *Odontophora* with teeth everted; (f) *Theristus*.

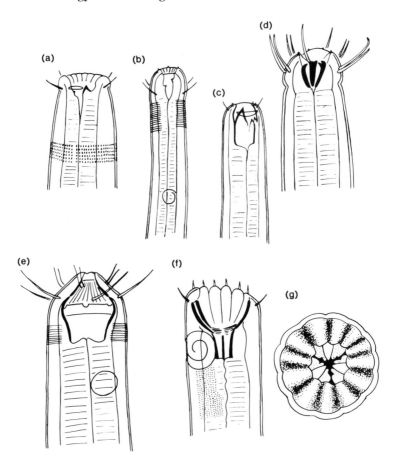

Fig. 8.3.   The association between the structure of the buccal cavity of marine nematodes and feeding category. Group 2A, epigrowth feeders: (a) *Dichromadora*; (b) *Calmicrolaimus*. Group 2B, omnivore/predators: (c) *Oncholaimus*; (d) *Enoplus*; predators (e) *Sphaerolaimus*; (f) *Gammanema*; (g) *Gammanema* head on.

1B. *Unselective deposit feeders*, 73 genera. These have cup-shaped conical or wide cylindrical buccal cavities, without teeth, which are flexible enough to engulf large particles, for example diatoms or crumbs of organic matter, often aided by the action of the lips.

2A. *Epigrowth feeders*, 104 genera. The buccal cavity is armed with teeth, rods, or plates. Food is scraped from surfaces, such as sand grains. Algal cells may be punctured or prized open.

2B. *Omnivor/predators*, 87 genera. Large buccal cavity with powerful teeth or plates. Prey may be swallowed whole and ruptured within the buccal cavity, or punctured and the contents sucked out.

Boucher (1973) has suggested a modified classification which subdivides two of the categories to take account of important differences in their food. This classification can be correlated to a degree with taxonomic sub-divisions

			Autumn (per cent)	Spring (per cent)
1A	*Selective deposit feeders*	Oxystomatidae ⎫ Leptosomatidae ⎭	1.9	3.3

1B	*Non-selective deposit feeders:*
	  (a) *Aggregate feeders*
	  (b) *Others*
		
		Comesomatidae
		Monhysteridae ⎫
		Axonolaimidae ⎭
			
			54.3
			11.2
				
				53.9
				11.3

2A	*Feeders on organic films* or *Epistrate feeders*	Desmodoridae ⎫ Chromadoridae ⎬ Cyatholaimidae ⎭	9.2	12.5

2B	*Omnivor/predators:*
	  (a) *Omnivorous*
	    particle feeders
		
		Enoplidae ⎫
		Rhabditolaimidae ⎬
		Ironidae ⎭
			
			14.4
				
				9.5

	(b) *Selective* *predators*	Halichoanolaimidae ⎫ Sphaerolaimidae ⎭	8.2	8.2

These figures give the percentage composition of the food categories in sublittoral muds from the Mediterranean Sea as an example (Boucher 1972–3).

In marine sediments nematodes are usually concentrated in the top 5–10 cm and many surveys have restricted their collections to these depths, but there are exceptions to this. In coarse sand on ocean beaches, for example, they may occur at much greater depths. Marine nematodes may extend well beyond the high-tide mark in saline coastal 'ground water' and some species are characteristically found in saline or brackish water above sea level beneath coastal dunes (Gerlach 1955).

Marine nematodes are, like terrestrial nematodes, affected directly and indirectly by the size and closeness of packing of mineral particles. A useful statistic is the median grain size, determined by standard sedimentological methods (Krumbein and Pettijohn 1938), though the frequency distribution of particle sizes and the closeness of packing, are also important. These properties are strongly influenced by wave action in coastal waters, so that the composition of the fauna can be related to the degree of exposure to wave action, as in Puget Sound, Wa. USA (Wieser 1959a). Wieser (1959b) studied relationships between grain size on the sandy beaches of Puget Sound, Wa. USA, and concluded that 200 μm may be the lower median grain size for nematodes to live in the interstitial spaces. Below 200 μm the nematodes may burrow rather than live in the interstitial spaces. Some marine nematodes are semi-sessile and tubiculous (Riemann 1974). Long cephalic and body setae are characteristic of nematodes living in sand, often exceeding 40 μm. Nematodes from muds, or very coarse deposits, have shorter setae, 5–10 μm (Gerlach 1953; Wieser 1959b; Warwick 1971). Ward (1975) found a correlation between the range of nematode lengths at a collecting station and the heterogeneity of particle sizes (though not with the median grain size). In poorly sorted sediments, with a wide mixture of particle sizes, the nematodes were more diverse in length. The elaborately sculptured cuticles of Ceramonematoidea, Monoposthoidea, and Xyalidae may aid in locomotion in coarse silt-free sediments in which these nematodes are common (Ward 1975).

Sublittoral marine sediments are usually sampled with an instrument which takes small cores, typically 10 cm^2 in an area 5 cm deep. These are likely to contain many individuals from several different families. Species composition and diversity can usually be correlated with the relative proportions of sands, silts, and clays making up the sediment. Soft muds, with high silt/clay percentages, have high densities, dominated by Comesomatidae (e.g. *Sabatieria*) and Linhomoeidae (e.g. *Terschellingia*), which are non-selective deposit feeders. In sands or muddy sands, epigrowth feeders are relatively more numerous, e.g. Chromadoridae, Desmodoridae, and Xyalidae. Some arbitrarily chosen examples of studies of sublittoral marine communities are referred to in Table 8.5. The references cited all give additional data on the physical properties of the sediments which I have not attempted to summarize. Many of the species have only been named to genus, because of taxonomic uncertainties, though recognized as species, with many genera represented by a single species. Sublittoral communities show seasonal variations in total population densities, though there is little evidence of synchronization in their reproductive cycles. Several generations are probably completed each year by most species (Hopper and Meyers 1967b; Juario 1975; Warwick and Buchanan 1971; Tietjen 1969).

Frequently several species from the same genus, differing only in minute morphological details, occur together, apparently contradicting the general principle that competition will exclude more than one species requiring the same resources in the absence of mitigating factors, such as a fluctuating environment or spacial heterogeneity. Small discontinuous resources, each patch sustaining only a single generation, favour co-existence. Closely related species may occupy different niches within the same small volume of sediment, for example, at different depths (Boucher 1973) or, despite the presence of similar mouth parts, select different food (Juario 1975). Even very small volumes of sediment, although superficially homogeneous, may provide a complex and diverse habitat for nematodes. Kito (1982) observed pairs of species of *Monhystera* and *Chromadora*

TABLE 8.5
Some studies of sublittoral nematode communities

Adjacent coast	No. of stations	Depths (m)	No. of species* at stations	Densities ($\times 10^3$ m^{-2})	Reference
Black Sea	Many	0–50	3–33	—	Filipjev (1921)
England, Irish Sea	3	20–166	—	1570–2515	Moore (1931)
Chile, 40–50°S	Many	5–400	40–67	—	Wieser (1959c)
New England, USA	3	12–18	63–103	789	Wieser (1960)
Florida, USA	1	<1	100	—	Hopper and Meyers (1967a, b)
New England, USA	4	—	54–76	1184–5163	Tietjen (1969)
England, North Sea	3	35–80	—	1570–2515	Warwick and Buchanan (1970)
France, Mediterranean Sea	1	35	50	2665	Boucher (1973)
England, Irish Sea	6	7–39	30–125	1610–565	Ward (1973)
France, Mediterranean Sea	Many	25–90	7–32	—	Vitiello (1974)
France, Mediterranean Sea	Many	2–6	14–28	—	Vitiello (1974)
Germany, North Sea	1	35	87	2867–5261	Juario (1975)
New England, USA	18	4.6–38.7	19–43	110–501	Tietjen (1977)

* In some of the works referred to species have been identified as to genus and not to species, but most of the genera are represented by a single species at one station.

which co-existed, but closer inspection revealed differences in food between members of the pair and seasonal differences in one pair in reproduction.

Nematode ecologists have studied the relationships between the abundance and diversity of nematode populations and the physical, chemical, and biological features of the environment, and considered how best to compare different communities. Many attempts have been made to find mathematical models with which to describe the data on abundance and diversity of species making up nematode communities. Poole's (1974) textbook on quantitative ecology discusses most of these models, though without reference to nematodes. Poole comments that 'by fitting models to the data, the abundance relationships of two or more communities can be compared. However, the models are heuristically motivated, and are of little value in determining the underlying interaction responsible for the observed abundance relationships among the species.'

For many purposes, the relative abundance of each species in a nematode community may be a useful statistic, usually expressed as a percentage. A histogram, showing the percentage abundance for each species in rank order, is sufficient to illustrate the degree to which a community is dominated by few or more species, as in Fig. 8.4. It is apparent that in silty muds a few species dominate the communities in

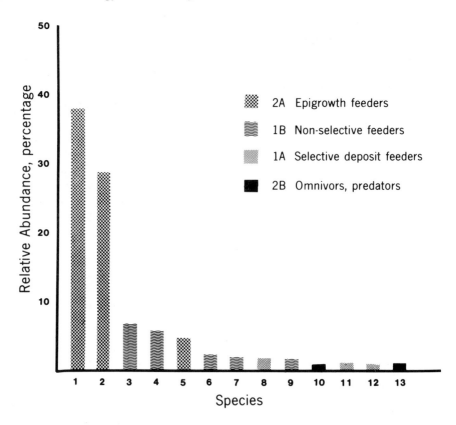

Fig. 8.4.   The relative abundance and food categories of species co-existing in a small area of estuarine mud flat (Candlagan Creek, NSW), showing the typical dominance of a few species.

terms of percentage abundance, characteristically belonging to the Comesomatidae (e.g. *Sabatiera* or *Dorylaimopis*) and Linhomoeidae (e.g. *Terschellingia*). In coarse sandy deposits, there are more relatively common species from a wider selection of families. Another general conclusion is that though the species present in different parts of the world differ, equivalent communities can be found with related species or genera dominating the fauna in physically similar environments. Warwick and Buchanan (1970) sampled three communities in the North Sea and found that both the logarithmic series and the log normal distribution fitted their data on relative abundance reasonably well. The logarithmic series gives a parameter α which is an index of diversity; for their communities 16.1, 15.3, and 22.5. Vitiello (1974) and Boucher (1973) used the same distribution for Mediterranean communities, ranging between 8 and 13 on different occasions. An extension of this has been to study the relationships of α to sample size (Ward 1973).

One method of comparing communities from different collecting stations, or levels, is to calculate the common abundance of each species. For example, if species A comprises 25, 15, and 5 per cent of the population at three stations, its common abundance at these three is the lowest percentage, 5 per cent. The sums of common abundances between different stations is a useful index for comparison which can be set out in a trellis diagram (see Fig. 8.5). The use of relative abundances has the disadvantage of failing to take into account differences in nematode densities. Alternatively, the number of species in common in a number of communities can be compared with a trellis diagram (Wieser 1960; King 1962; Hopper and Meyers 1967a; Tietjen 1969, 1971; Vitiello 1974). Juario

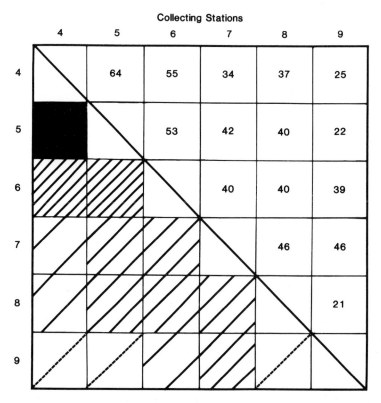

Fig. 8.5. The percentage common abundances of species of nematodes inhabiting six numbered collecting sites forming a transect of an estuarine mud flat (Hunter River, NSW). The percentage common abundances shared by each pair of sites is shown in the form of a trellis diagram, with their magnitude also indicated by the degree of cross-hatching.

(1975) used a different index of affinity, $Q = 2j/a + b$ where j is the number of species in both places and a and b the number restricted to one or the other, respectively. A mathematically more precise alternative to the trellis diagram for making comparisons is to use a computer program to estimate degrees of affinity and produce a dendogram (Tietjen 1976, 1977; Ward 1973). All these studies have emphasized the importance of the coarseness or fineness of the particles making up the sediment in determining the fauna.

Many investigators have applied statistics from 'information theory' to analyse species diversity. The number of species present is a measure of diversity, but does not take into account the relative abundance of the species present. The Shannon–Wiener function H' has been the most widely used (Juario 1975; King 1962; Vitiello 1974; Tietjen 1977; Dinet and Vivier 1979; Heip and Decraemer 1974). It assumes that a collection is a random sample from a population in which every species in the community is represented (though the accidental omission of several rare species does not lead to significant errors in estimating H'). It estimates the uncertainty of correctly predicting the species of an individual when one is drawn at random from the population.

$$H' = \sum_{i=1}^{s} P_i \log_2 P_i$$

where s is the number of species and P_i the proportion belonging to the ith species. The maximum possible value of H for a community is given by

$$H_{max} = \log_2 s.$$

Evenness J (or E), also widely used, measures the evenness with which the numbers of individuals is distributed amongst the species present,

$$J = H'/H_{max}.$$

The biological meaning of these parameters in describing nematode communities is uncertain. A graphical representation of the distribution of abundances on species in rank order illustrates the same data. However, since there are appropriate tests of significance for H' and J, these statistics may permit comparisons between communities to be made with greater confidence.

Values for H' vary widely in individual collections, i.e. 1.56–10.5, but are lower in finer sediments than coarse ones (Vitiello 1974; Juario 1975; Heip and Decraemer 1974; Dinet and Vivier 1979; Tietjen 1976, 1977). Evenness, J, also varies widely from 0.51 to 0.96. Hiep and Decraemer (1974) found strong linear correlations between evenness and diversity and median grain size in coastal communities off the Belgian coast. All were negatively correlated with the percentage of fine silt in the sediments sampled. Tietjen (1977) also found a strong positive correlation between diversity and evenness in communities from New York Sound. Investigators have attempted to use indices of diversity as measures of the effects of pollution and Tietjen (1980*b*) observed that organic carbon and heavy metals probably depressed diversity in medium sands in New York Bight, but not in silty sands.

Less is known about the fauna of deep oceanic sediments. Tietjen (1971, 1976) took two transects extending from the coast of North Carolina into the Atlantic across the continental slope. The population density generally declined with depth from 50 to 2500 m, varying with the nature of the sediment. Population densities varied between 32 and 1026 $\times$ 10^3 individuals m^{-2}. Approximately 50 per cent of the 206 species were restricted to one of the four types of sediments recognized, with 24 to 30 genera present at each station. Epigrowth feeders declined sharply below 500 m, where deposit feeders became relatively much more numerous (assuming Wieser's categories are applicable to deepwater sediments). Species diversity, by numbers of species and by H', and species richness declined with depth, while evenness J remained uniformly high at all depths, 0.82–0.96. Different results were obtained by Dinet and Vivier (1977, 1979) in a study of the nematodes from the Bay of Biscay at six stations between 1920 and 4725 m. The abyssal plain was evidently very rich in species, 317 being found, though dominated by one genus; *Theristus* (*sensu lato*) (*Xyalidae*). Diversity, as measured by H', was moderate (5.24–6.67) and evenness uniformly high (0.93–0.96).

In the littoral zone, wave action and tidal influences may exert much greater effects than in the sublittoral zone. These determine the coarseness of the sediments and their degree of sorting and strongly influence the quantity of organic matter and penetration of oxygen. Consequently the composition of the nematode fauna is greatly influenced by the intensity of wave action, with the median grain size being a useful statistic. Ocean beaches have a rich fauna of nematodes. Gerlach (1953), for example, recorded 188 different species from North German shores, and other investigators have found similar diversity in species, genera, and families. However, only a small fraction of these species occur in any one sample; the fauna being divisible into many separate communities, most readily correlated with the coarseness of the sediment (Wieser 1959*a*, *b*, *c*). It is possible to recognize at one locality, species with restricted distributions on beaches (stenotopic) and species with widespread distributions (eurytopic). The littoral meiofauna resembles the sublittoral fauna in many respects, though perhaps generally less dense, but may extend to greater depths in sand on exposed beaches, and show zonation imposed by differing periods of exposure at low tide to the atmosphere.

Some species are found on beaches exposed to very heavy surf, notably some Enoplidae and Xyalidae. How they retain their position is puzzling. On an Atlantic beach in Scotland, McIntyre and Murison (1973) observed that some nematodes moved deeper into the sand as the waves passed over the region they occupied, with relatively few nematodes present in the surf. In South Bay on the Northern Ireland coast, which I presume to be more sheltered, there was relatively little vertical migration as the waves passed over the nematode community (Boaden and Platt 1971). Wave action circulates water through the surface layers of sand oxygenating the environment, but deoxygenation when the tide is out may induce migration (Boaden and Platt 1971). I have found populations which persist on Australian beaches even after winter storms have removed sand to a depth of one to two metres. They are probably re-distributed by the surf at each tide. Galtsova (1981) observed seasonal changes in the vertical distribution of nematodes in a sandy beach on the White Sea in summer and autumn, primarily correlated with temperature.

Even within physically uniform sands, some species may show irregular distribution (Findlay 1981) while others appear more uniformly dispersed. Aggregation in some species may be caused by attraction to food. Gerlach (1977) found some species apparently attracted to decomposing fish on a Bermuda beach and Meyers, Hopper and Cefalu (1970) found dense aggregations of *Metoncholaimus scissus*, reaching 2.68×10^6 m^{-2}, at the edge of turtle grass beds off the Florida coast. This large oncholaimid is attracted by fungal cellulose traps (Meyers and Hopper 1966), though its actual diet is unknown. Very dense communities may occur on some beaches, with Gerlach (1953), for example, reporting 5×10^6 nematodes m^{-2} in sand rich in Cyanophyceae.

The macroscopic algae growing on rocks and other hard surfaces support characteristic nematode communities in the intertidal and shallow sub-littoral zone (Filipjev 1921; Wieser 1959a; Warwick 1977; Moore 1971; Kito 1982). Very similar populations occur on sessile animals. Ocelli or eye spots are common on algal (or phytal) nematodes, especially those feeding on the epigrowth of microscopic algae and bacteria. *Chromadorita tenuis* is chemically attracted and swims to macrophytes (Jensen 1981). The form of the algae determines to a large extent the species composition found on the algae, larger nematodes occurring in larger fronds. Seasonal fluctuations are greater in phytal nematodes than in those from sand and mud. In Japan, numbers are greatest in spring and autumn, with seasonal variations in species composition and morphology throughout the year (Kito 1982). A characteristic fauna of large omnivor/predators is associated with the holdfasts of large brown algae. Finely branching algae retain fine silt, so that deposit-feeding nematodes, such as *Theristus*, which are also common in fine silt elsewhere, are found in these collections of silt. Shallow water beds of sea grass (four families of Monocotyledons) support populations of nematodes which feed on epigrowths of algae, fungi, and bacteria, as well as giving rise to sediments very rich in organic matter and nematodes (Hopper and Meyers 1967a, b).

With the exception of exposed beaches, bacterial decomposition of organic matter gives rise to an anaerobic environment, rich in H_2S, below the region oxygenated by circulating sea water (Fenchel and Riedl 1970). The controversial existence of a fauna of nematodes and other meiobenthos adapted to this environment, a thiobios, has already been discussed (see Section 3.7). The existence of such an anaerobic thiobios fauna has been doubted by Reiser and Ax (1979). Most of the intertidal nematodes in muddy environments are concentrated in the upper 5–20 cm (though nematodes occur at much greater depths in clean sand). The vertical zonation of reducing conditions and the chemical properties of sandy beaches in the Irish Sea have been studied by Boaden and Platt (1971) and McLachlan (1978). Reducing conditions approach to within 5–10 cm of the surface on sandy beaches, with the depth of the sharp discontinuity in physico-chemical conditions, the redox-potential-discontinuity, or RPD, varying with temperature, time of day, and water circulation. Seasonal changes in the vertical distribution of nematodes in Strangford Lough, a sheltered bay in Northern Ireland, are probably due to the seasonal variations in the depth of the RPD (Platt 1977b). On sandy beaches nematodes also found in sublittoral coarse bottoms may extend into the littoral zone in deep 'ground water' (Gerlach 1953, 1955, 1967).

Very dense populations of nematodes occur in estuarine mud flats (Warwick and Price 1979; Skoolmun and Gerlach 1971), mangrove swamps (Hopper et al. 1973), and salt marshes (Wieser and Kanwisher 1961; Teal and Wieser 1966). In such places the RPD almost reaches the surface of the mud, but nematodes are numerous in the anaerobic mud 5–20 cm below the surface. In the Lynher estuary in Cornwall, England, the population of about 40 species varied from 8 to 23 × 10^6 m^{-2} in May (Warwick and Price 1979). Seasonal changes occur in the total population, usually reaching maxima in summer, but there is little synchronization in reproductive cycles, and different species reach their maximum at all times of the year (McIntyre and Murison 1973; Skoolmun and Gerlach 1971). *Oncholaimus oxyuris* is a brackish-water species which achieves high densities of 40–50 × 10^3 m^{-2}, but with low reproductive potential. One or two generations occur each year, with gravid females present from spring to early summer (Smol, Hiep, and Govaert 1980).

Marine nematodes may extend far up the tidal reaches of river estuaries, with such typical mud-living genera as *Sabatieria*, *Spirinia*, and *Viscosia* represented. Factors in addition to salinity are important in determining the fauna in British estuaries (Capstick 1959; Warwick 1971). In the Baltic

Sea typical marine nematode communities are found with average salinities as low as 12.5‰ (Brenning 1973). Nematodes associated with brackish water; between 3 and 30‰, have been described from the Baltic Sea and the Zuider Zee (Filipjev 1929–30; Gerlach 1953). Gerlach (1953) found the lowest population at 5‰ in the Baltic, and Jensen (1979a) found few nematodes in rock pools off the Finnish coast.

The relative importance of nematodes in marine ecosystems has been discussed by McIntyre (1968), Gerlach (1971), and Platt and Warwick (1980). Platt and Warwick conclude that densities are highest in muddy estuaries and salt marshes, ranging from 10 to 23×10^6 m^{-2}. Nematodes are a major component of the biomass. Surprisingly, little is known about the importance of nematodes in the diet of other marine animals. Circumstantial evidence has suggested that some polychaetes may feed on nematodes. Shrimps are probably important predators (Bell and Coull 1978). Detritus feeding crustacea, molluscs, juvenile fish, aquatic birds, and even elephant seals may ingest quantities of meiofauna, but whether the nematodes are digested or contribute significantly to their diet remains to be discovered. Table 8.6 lists references to studies of intertidal nematode communities from a wide variety of habitats.

8.7. Bacteria-feeding by nematodes

Bacteria-feeding nematodes are so common in many ecosystems that it is important to assess their importance in the decomposition of organic matter and the recycling of nutrients and to understand their population dynamics. In the soil many species coexist, which in the laboratory will feed on a wide range of bacteria under similar conditions (see Section 4.2), but may adopt different strategies in nature. Anderson and Coleman (1981) have analysed the different life-cycle strategies adopted by two widespread bacteria-feeding nematodes, collecting both species from prairie soil in Colorado. *Acrobeloides* sp does better with lower bacterial densities than *Mesodiplogaster lheritieri*. *M. lheritieri* reproduced rapidly (generation time 4 days) completing egg production within 10 days in laboratory culture. It responded to food shortage by producing dauer larvae or intrauterine matricidal larvae. *Acrobeloides* grew more slowly (generation time 11 days), sustained egg production for longer, and responded to adverse conditions by forming quiescent larvae and adults which have greater powers of survival. In competition *M. lheritieri* may give rise to predaceous forms which feed on *Acrobeloides*. Both species are attracted to potentially useful food, *Pseudomonas*, *Arthrobacter*, and *Escherichia coli*.

Of particular interest are life-tables drawn up for both species in culture. In addition to giving information on growth, reproduction, and mortality, a

TABLE 8.6
Some studies of the intertidal nematode fauna

Locality	Habitats	Reference
Black Sea, USSR	Sand, mud, rocks, algae, etc.	Filipjev (1921)
North Sea & Baltic Sea, N. Germany	Sandy and silty beaches	Gerlach (1953)
Puget Sound, USA	Sandy beaches	Wieser (1959a)
Chile, 40–50°S	Sandy, silty, muddy beaches, algae	Wieser (1959c)
Florida, USA	Sandy beaches	King (1962)
Baltic Sea, N. Germany	Sand	Brenning (1973)
South Bay, N. Ireland	Sand	Boaden and Platt (1971)
Strangford Lough, N. Ireland	Sand	Platt (1977a, b)
Vancouver Island, Brit. Columbia	Sand and muddy beaches	Sharma, Hopper and Webster (1977, 1978)
White Sea, USSR	Rocky, sandy beach	Galtsova (1976)
English Channel	Algae	Wieser (1952)
North Sea, England	Algae, holdfasts	Moore (1971)
Scilly Islands, England	Algae	Warwick (1977)
Hokkaido, Japan	Algae	Kito (1982)
Massachusetts, USA	Salt marsh	Wieser and Kanwisher (1961)
North England	Estuarine mud	Capstick (1959)
Georgia, USA	Salt marsh	Teal and Wieser (1966)
South England	Estuarine mud	Warwick (1971)
Florida, USA	Mangrove swamp	Hopper, Fell, and Cefalu (1973)
Great Barrier Reef, Australia	Mangrove swamp	Decraemer and Coomans (1978)
South England	Estuarine mud	Warwick and Price (1979)

number of useful population parameters can be calculated from such tables (Poole 1974), for example, the net reproductive rate R_0, and the intrinsic rate of increase.

Many bacteria-feeding soil nematodes are r-strategists, with short generation times and high fecundity (see Table 8.3), capable of rapidly exploiting transient rich resources in the form of decomposing organic matter. In comparing the way different species exploit such potential resources their intrinsic rate of increase r may be a more useful statistic than their generation time and fecundity. Life-tables measure the growth rate, average number of offspring per female (or hermaphrodite), and the mortality at each stage of growth. The number of males is not important when there are sufficient to fertilize all the eggs produced; or with self-fertilizing hermaphroditic or parthenogenetic species. Many bacteria-feeding nematodes are self-fertilizing hermaphrodites. When resources are not limiting, the numbers of individuals will increase exponentially with time at a rate which is a function of the number of females at any given moment, but the relative rate of increase r over a constant time interval will remain the same, while conditions remain the same, and will be associated with a stable population structure.

Life-tables were constructed for *Acrobeloides* sp and *M. lheritieri* by Anderson and Coleman (1981) from which they calculated r and the net reproductive rate R_o (or multiplication per mean generation time) as well as other useful statistics. Mortality at each stage of growth was estimated, information not generally available for other species. For *Acrobeloides* $r =$ 0.34 and $R_o = 40.1$; for *M. lheritieri*, $r = 0.88$ and $R_o = 33.8$. Anderson and Coleman (1982) describe the mathematical model simulating the population dynamics of these two species, which they use to predict carbon flows when the nematodes feed on bacteria.

The intrinsic rate of increase can also be calculated less precisely from population growth over longer time intervals from the formula

$$r = \frac{1}{t} \ln \frac{N_t}{N_o}$$

where N_o is the initial population and N_t the population at time t. The value of r is strongly influenced by temperature in *M. lheritieri* (Grootaert 1976). Alongi and Tietjen (1980) calculated r for several salt marsh nematodes feeding on bacteria at 23 °C in this way. For *Diplolaimella* sp values of $r =$ 0.096–0.086 were obtained; for *Monhystera disjuncta*, 0.099–0.093; for *Chromadorina germanica*, 0.064–0.068. They differed in their ability to feed on several algae, with corresponding differences in r; and when two species of nematode were cultured together on bacteria, r was depressed. Interactions between algae and bacteria and mixed cultures of nematodes depended on whether they competed for food or not. Warwick (1981) found a much higher value, $r = 0.241$, for another salt-marsh species *Diplolaimelloides brucei* at 23 °C and 26‰. In this species, temperature had a strong effect on r, salinity less so. Data on the growth of terrestrial nematode populations on bacteria can also be found in Yeates (1970).

The extent to which the maximum intrinsic rate of increase is sustained must be influenced by the density of bacterial food required for maximal growth. *C. briggsae* requires dense suspensions of *Escherichia coli* to sustain growth. Nicholas, Grassia, and Viswanathan (1974) found 6.28×10^8/ml would maintain population increase, 6.28×10^7 allowed survival, but at 6.28×10^6 the nematodes died. The rate of digestion was a function of bacterial density. When a population of *C. briggsae* was allowed to feed on a dense suspension, the numbers of bacteria fell exponentially with time to a minimal value below that necessary for survival

$$Y = ae^{-bt},$$

where a was the initial density, b a function of the feeding rate, and t time. This model implies that the rate of feeding remained constant as bacterial density declined from 7×10^9 to 5×10^8 cells/ml.

It seems that *C. briggsae* requires dense populations of bacteria for growth. In three experiments lasting for 96 hours (Nicholas *et al*. 1974), in which a population of *C. briggsae* fed upon a suspension of *E. coli*, the increase in dry weight of the nematode averaged 13 per cent of the weight of bacteria consumed (range 10.4–14.8 per cent). In a long-term culture the figure was 9 per cent, but no account was taken of nematode mortality. A few larval nematodes were introduced to 4.5 ml of bacterial suspension (6.28×10^9 ml^{-1}) and allowed to multiply (beginning on day five) for 27 days before the experiment was terminated.

Several estimates have been made for the numbers of bacteria consumed. Mercer and Cairns (1973) estimated indirectly that females of *Pelodera chitwoodi* consumed 10×10^6 bacteria in eight days, males 8.1×10^6. A reproductive female of *Plectus palustris*, a freshwater species, feeds continuously on *Acinobacter* sp at densities of $5–10 \times 10^9$ bacteria/ml, with a reproductive female taking about 7.23×10^6 cells, or 1.94 µg dry weight per day. Tietjen and Lee (1977*b*) used ^{32}P-labelled bacteria and algae to estimate consumption by salt-marsh nematodes: *Monhystera denticulata* consumed 5.7×10^{-2} µg/nematode/day, *Rhabditis marina* 1.8×10^{-2} µg/nematode/day. Their data show significant differences in the rate with which several nematodes feed on bacteria and various algae in the laboratory.

Far too little is known about many factors which influence the rate of consumption to usefully extrapolate from such examples to the effects of nematodes on bacterial populations under natural conditions. Bacteria-feeding and predatory nematodes reach high densities in sewage treatment plants. Calculations suggest their direct effects on the catabolism of organic matter may be minimal, but that they may have a significant effect through stimulating bacterial multiplication as a consequence of their consumption of bacteria (Schiemer 1975).

Experiments on the interactions of nematodes, amoebae, and bacteria under laboratory conditions have been reported by a group of workers at Colorado, USA, in what are termed microcosm studies. Soil after sterilization with propylene oxide, was inoculated with combinations of bacteria, *Pseudomonas cepacea*; amoebae, *Acanthamoeba polyphaga*; and the nematode, *Mesodiplogaster lheritieri*; and the populations monitored for many days. Soil, appropriately moistened, was contained in Erlermeyer flasks and sometimes supplemented with glucose. Both nematodes and amoebae fed on the bacteria, significantly reducing the bacterial population, but when both were present together, the nematode population reached higher densities by feeding on the amoebae. The addition of glucose stimulated population growth by all three. When food became scarce, the dauer larvae accumulated in place of adult nematodes (Anderson, Elliott, McClellan, Coleman, Cole, and Hunt 1978).

When respiratory CO_2 evolved from the microcosms was monitored (Coleman, Anderson, Cole, Elliott, Woods, and Campion 1978), both amoebae and nematodes increased CO_2 production compared with bacteria alone. The addition of glucose stimulated respiration in all combinations, but there were differences in the partitioning of added carbon between CO_2 and the biomass. The amoebae accumulated much more carbon while the nematodes respired 30 times more per unit biomass. Bacteria on their own progressively assimilated inorganic phosphorus from the soil, which, when amoebae were present without nematodes, was returned to the pool of labile inorganic phosphorus (Cole, Elliott, Hart, and Coleman 1978). Nematodes, without amoebae, had little effect on bacterial phosphorus, but with amoebae the effects were not clear. The authors conclude that nematodes accumulate a significant part of the phosphorus within themselves.

Microcosm experiments with potted pine seedlings growing in humus/ sand-mixtures have been used by Swedish workers to investigate the effects of stimulating microbial activity on the growth of pines, especially by releasing nitrogen from the humus (Baäth, Lohm, Lundgren, Rosswall, Söderström, Sohlenius, and Wirén 1978). Adding nitrogen stimulated population increases in bacteria-feeding nematodes (e.g. *Rhabditis* sp and *Diplogaster* sp); adding glucose increased the number of *Aphelenchoides*, probably by stimulating fungi; while *Acrobeloides* was commonest in the control pots (a bacteria-feeder).

When a population of *C. briggsae* feeds on *E. coli* previously labelled with $^{32}PO_4$ or [^{14}C]NaHCO$_3$, about 20 per cent of the label becomes incorporated into the nematode tissues, while much of the rest appears in the medium as secretions or excretions. About 14 per cent of ^{14}C is incorporated into high molecular weight compounds, with much of the rest lost as ^{14}[C]O$_2$ (Nicholas and Viswanathan 1975). With ^{32}P the nematodes rapidly exchange the label with the culture medium. After ^{14}C has been incorporated into the tissues, by feeding on ^{14}C-labelled *E. coli*, the label is progressively lost over the following 10 days, if feeding on labelled bacteria is discontinued. Nicholas and Viswanathan (1975) measured the distribution of both radioactive labels within biochemical fractions of nematode tissues. An important observation is that a much higher proportion of the ^{14}C-label is retained in the nematodes' tissues when the *E. coli* were pre-labelled with [^{14}C]-glucose or [^{14}C]-amino acids than with [^{14}C]-NaHCO$_3$, indicating preferential retention of these molecules by both organisms.

The common observation that several bacteria-feeding nematodes occur together in the soil, each capable of feeding on a similar range of bacteria in the laboratory, raises questions about competitive exclusion. Anderson and Coleman's (1981) study of *M. lheritieri* and *Acrobeloides* shows that though these species may compete, differences in their biology imply that each will have the advantage under different environmental circumstances. However, mathematical simulations of competing species suggest that prolonged co-existence is possible between a number of species sharing the same resource, when this is small, isolated, and likely to support only a single generation, or when individuals of different species aggregate independently over isolated limited resources (Atkinson and Sharrocks 1981). Nematodes are attracted to bacterial food and it may be that both these criteria are applicable to the co-existence of some bacteria-feeding species.

8.8. Bioenergetics

The great numbers of nematodes present in all kinds of ecosystems has already been stressed. They are amongst the most numerous of animals feeding on the primary decomposers, bacteria, actinomycetes, and fungi, whilst others are important in feeding on the primary producers, algae and higher plants. Nematodes provide food for other animals and fungi as well as recycling plant nutrients and carbon through respiration. A difficult question to answer is what is the significance of their activity in the overall economy of the ecosystem? One yardstick is the relative contribution made by nematodes to the total flow of energy through the ecosystem. Though this is only one of many ways in which their importance might be established, it has the advantage that it makes possible a direct comparison between nematodes and other animals. Its disadvantages are that it is difficult to sum the effects of all the species, at all stages of their life-cycles, without prodigious efforts at counting, measuring, and sampling, and that the calculations must also use laboratory measurements, which must be made under conditions which may not satisfactorily represent conditions in nature. Nonetheless, several attempts have been made to estimate the contributions of nematodes to energy flow through various terrestrial and marine ecosystems.

Estimating the flux of energy through a population ideally requires evaluation of the energy content of each term in the equation

$$C = P + R + E + U$$

over a stated interval, usually a year, where C is consumption, P is production, R is respiration, E is ejecta, and U is excreta. The preferred SI unit is the kilojoule, but until recently was the calorie (1 calorie = 4.186 J).

Production requires estimates of the numbers, growth rate, and reproduction of all the species present in significant numbers, taking into account seasonal changes in temperature. The estimates are first made in weight or biomass and then converted to kilojoules (or calories) from measurements of calorific value per unit weight. The energy liberated by respiration is measured in the laboratory (see Section 3.7). Laboratory experiments are also required to estimate the rate of consumption, efficiency of assimilation, and, usually by difference, the proportion of consumption appearing as ejecta (i.e. faeces and excretion). The task would prove impractical were it not possible to derive some general approximations so that these estimates need not be based on measurements on every species present. For example, several workers have used a general formula to estimate the production of nematodes from their respiration. These depend on equations which relate production and assimilation to community respiration by invertebrates generally (McNeil and Lawton 1970). It may be possible to generalize about production and assimilation for short-lived aquatic invertebrates from biomass and weight per unit area (Humphreys 1981), but this has yet to be applied to nematodes.

Pioneering attempts to estimate the contribution of nematodes to energy flow in terrestrial ecosystems were made by Overgaard Nielsen (1949) for Danish soils, and by Teal (1962) for a marine salt marsh. However, much better estimates have now been made for both kinds of habitat. Sohlenius (1980) and Yeates (1979b) have critically reviewed more recent estimates for terrestrial ecosystems. It is probably premature to make any definite statement about the relative importance of nematodes in general. Estimates for their contribution to terrestrial soil respiration have varied rather widely from about 0.3 to 1 per cent. Sohlenius believes they could account for 10–15 per cent of the animal soil respiration. In a later paper, Caldwell estimates that during the short summer on an Antarctic Island, Signy Island, moss nematodes account for 16–25 per cent of the metazoan respiration. Using McNeil and Lawton's (1970) equation for short-lived poikilotherms, this would be equivalent to a production of 0.95–1.12 kCal m^{-2} y^{-1}.

I would like to refer to two of the more ambitious and comprehensive attempts to estimate the nematode energy flow. Yeates (1973a, b) studied the nematode population in a Danish beech forest, making the following estimates, based on 29 species sampled at monthly intervals

Biomass, annual mean = 370 mg m^{-2} (1 432 000 individuals m^{-2});
Production = 1.5–3 mg m^{-2} yr^{-1} (4.06 × mean monthly biomass)
= 3.230 kCal m^{-2};
Respiration = 1091 ml O$_2$ m^{-2} yr^{-1}
= 5.21 kCal m^{-2} yr^{-1}

Biomass was converted to calories from Yeates's (1972b) measurements by bomb calorimetry giving 2.152 Cal mg^{-1}. A figure of 4.775 Cal mg^{-1} was used for converting O_2 consumption to its calorific equivalent.

Warwick and Price (1979) studied the nematode fauna of the Lynher estuary which empties into the Tamar River estuary in Southern Britain. Their work gains in significance because comparable studies are being made by them and associated workers on other organisms in the same estuary, leading to a comprehensive budget for the ecosystem (Warwick, Joint, and Radford 1979). About 40 nematode species are present on the mud flat with a population ranging from 8 to 9 × 10^6 individuals m^{-2} (=1.4–1.6 g m^{-2}) in winter to a peak of 22.86 × 10^6 (=3.4 g m^{-2}) in May. Studies of the respiration of the 16 commonest species formed the basis of their work (see also Section 3.7). They calculate a total oxygen consumption of 28.0 l m^{-2} y^{-1}, equivalent to 11.2 g carbon metabolized. Annual production, calculated from respiration according to McNeil and Lawton's (1970) equation, would be 6.623 g carbon m^{-2} y^{-1}. The Lynher estuary possesses a very dense population of nematodes, when compared with other marine and terrestrial habitats which have been sampled with comparable thoroughness, which is equivalent to 15 per cent of the macrofauna. Moreover, they continue to multiply throughout the year. The values for community respiration, estimated by Wieser and Kanwisher (1961) and Teal and Weiser (1966) for nematodes in salt marshes on the Atlantic coast of America, are rather close when adjusted for different temperatures. Warwick and Price suggest that a mean value of 6 l O_2 m^{-2} y^{-1} g^{-2} might be a useful approximation with which to estimate respiration from biomass, after correction for differences in temperature.

Amongst the laboratory measurements required for calculating energy budgets are the calorific equivalents of wet or dry weights. These have been measured by bomb calorimetry (Paine 1971) (see Table 8.7). Some assorted values are also given by Croll and de Soyza (1980).

TABLE 8.7

Measurements of the calorific value of nematodes by bomb calorimetry

Species	Food	Energy g^{-1} ash-free dry weight		Reference
		(kJ)	(kCal)	
Aphelenchus avenae	fungi	22.83	5.453	Soyza (1973)
Plectus sp. + *Poikilolaimus* sp.	bacteria	17.94	4.285	Yeates (1972b)
Rhabditis oxycerca				
(= *Pelodera* sp.)	bacteria	26.4	6.300	Marchant and Nicholas (1974)
Caenorhabditis elegans	bacteria	26.45	6.316	Nicholas and Stewart (1978)

Several energy budgets have been drawn up for single species from laboratory measurements. They are important for confirming the validity of the equations used to calculate community energy budgets, but they are difficult to summarize because they are drawn up on different bases. I have summarized several below, converting calories to joules where necessary. They give energy conversions per day. Cumulative budgets have also been calculated for the life-span and egg production. A female of *A. avenae* consumed 44 mJ, used 8.0 mJ for growth, and used 17 mJ for eggs in 25 days on a fungal diet. A *Pelodera chitwoodi* male consumed 20 mJ of bacteria, a female 67mJ, in a life-span of eight days.

	Energy budgets*; ($J \times 10^{-3} d^{-1}$)		
Species	*Aphelenchus avenae***	*Plectus palustris*†	*Rhabditis oxycerca*‡
basis	adult ♀	reproductive ♀ (1.5 µg)	mixed population; per µg
food	fungi	bacteria	bacteria
consumption	2.18	40.7	3.7
assimilation	—	5.03	—
production	—	4.12	0.83
—growth	1.70	0.55	—
—eggs	1.62	3.56	—
ejecta	—	—	1.5
respiration	1.65	0.92	1.37

* Calories converted to joules by multiplying by 4.186.
** Soyza (1973).
† Duncan, Schiemer, and Klekowski (1974).
‡ Marchant and Nicholas (1974).

Perhaps more significant are estimates of the ecological efficiencies (see Table 8.8). For this purpose Tietjen's carbon budgets can also be included, though a direct conversion from carbon to energy flow is not possible without data on composition. The estimates of production efficiency are very high compared with those accepted for other animals, but caution must be exercised in their interpretation. The relative proportions of energy diverted to respiration and production will be strongly influenced by food density, since the work done in finding food will be important (Marchant and Nicholas 1974). With *R. oxycerca* Marchant and Nicholas (1974) calculated production efficiency at 38 per cent. Similar experiments with *C. briggsae*, in both cases measuring the partition of radioactive carbon after feeding on labelled bacteria gave a value of 20 per cent (Nicholas and Viswanathan 1975). Gravimetric measurements of the conversion of bacterial food into nematode tissue under similar experi-

TABLE 8.8

Ecological efficiencies; percentages

	A/I	P/I	P/A	R/A	Reference
Aquatic bacteria-feeders					
Rhabditis oxycerca	60	22	28	62	Marchant and Nicholas (1974)
Plectus palustris	12	10	82	18	Duncan *et al.* (1974)
Marine bacteria-feeder					
Diplolaimelloides brucei	—	—	86.9	18	Warwick (1981)
Marine algal-feeder					
Chromadorina germanica	6.2	4.9	14.6	25.0	Tietjen (1980*a*)
Marine bacteria-feeder					
Monhystera disjuncta	18.3	14.6	79.8	20.2	Tietjen (1980*a*)
Rhabditis marina	25.8	25.0	96.5	3.5	Tietjen (1980*a*)
Terrestrial fungal-feeder					
Aphelenchus avenae	—	—	67	37	Soyza (1973)

A = assimilation; I = ingestion; P = production; and R = respiration.

mental conditions gave 13 per cent (Nicholas, Grassia, and Viswanathan 1974).

Warwick's (1981) data in *D. brucei* are instructive. Productive efficiency $P/P+R$ rose from around 40 per cent at the annual minimum temperature of estuarine water to 87 per cent at 15 °C, falling slightly at higher temperatures. Salinity also affected the efficiency. Reproductive effort remained fairly stable at about 10 per cent of assimilation (approximately $P + R$).

9. Taxonomy

9.1. Introduction

NEMATODE taxomony is in a state of flux, with the publication of several different classifications in the last few years, while at a lower taxonomic level, the description of many new species has led to continual rearrangement and subdivision of taxa. Most nematologists, in common with the majority of zoologists, believe that the classification of animals should be based on their phylogeny, and all the recent classifications have sought to do this, though here we are on very uncertain ground. This principle has not gone unchallenged, and numerical taxonomists, who might more accurately be described as phenetic numerical taxonomists, believe that any classification should begin by comparing a wide range of characters, without reference to, in the first instance, phylogenetic theory. Their methods have been applied to the lower taxonomic divisions (Bird and Mai 1967; Moss and Webster 1970; Blackith and Blackith 1976) but have not had much influence on nematode classifications.

The use of mathematical methods to analyse taxonomic data does not necessarily mean that the phylogenetic considerations must be set aside, and Maggenti (1970), who used systems analysis to explore the phylogeny of nematodes, came up with propositions similar to those advanced by Andrassy (1976) in major taxonomic work on more conventional lines. The cladistic school of systematists, which has its strongest following among German zoologists, concentrates instead on reconstructing the phylogenetic history of a group of animals from their comparative morphology, focusing its attention on the actual branch points in the phylogenetic tree. A very useful discussion of the general principles of systematics as they apply to nematodes can be found in an article by Coomans (1979a).

Whatever guiding principles the nematode taxonomist follows in classification, the Nematoda presents great difficulties. Though there are a large number of species, the majority are remarkably similar morphologically. This similarity, it has been suggested, comes from a similar locomotory and skeletal system. Consequently, the taxonomist must concentrate on structures associated with the mouth parts, the sense organs, the cuticle, and the copulatory apparatus. Whichever classification is adopted, convergent evolution seems to have been common. The concept of homology is of crucial importance for making appropriate comparisons, but the criteria which can be used to establish homologies are difficult to apply to such structurally simple animals. Nematodes are so small that the taxonomically useful characters are often at the limit of the resolving power of the light microscope. Scanning electron microscopy, and to a lesser extent transmission electron microscopy, have proved invaluable for revealing the structure of minute characters, but both are destructive methods so that the specimens once examined can no longer form part of a type collection. The few nematode fossils are of recent forms and of no assistance to the nematode taxonomist.

Such considerations primarily affect the higher taxonomic categories and, at the family level or below, pragmatism has been more important than theory. Nematode taxonomists concerned with medical, veterinary, or plant parasites urgently require reliable methods for identification. As great numbers of species continue to be identified, there has been a tendency to repeatedly subdivide taxa, raising the ranks of the taxa involved. Animal parasitologists, terrestrial nematologists, and marine nematologists have worked largely independently of one another, so that the continual subdivision and raising of taxonomic categories has thrown the whole classification into imbalance. None of the recent classifications have been comprehensive, so that the last truly comprehensive classification is that published in *Traité de zoologie* (Grassé 1965).

Many authors have speculated on the phylogenetic history of nematodes and have attempted to recognize which nematode taxa are most primitive. In the absence of useful fossils, other evidence for the retention of a number of primitive morphological features by a nematode taxon would be of great value in deriving a phylogenetic classification. Riemann (1977*a*) has critically analysed the evidence from morphology and ontogeny which can be brought to bear on this problem. He suggests that the Diphtherophoroidea, Trichodoroidea, Onchulinae, and the genus *Tripyla* may be the most primitive taxa. Most nematodes depend on a hydraulic skeleton for locomotion, and its poor development in these nematodes may well be a primitive feature. Other authors have come to a similar conclusion about the primitive nature of these taxa on other evidence.

There is much still to be learned about nematode phylogeny but new insights may have to come from the application of molecular biology to the problem. The determination of the amino-acid sequences in commonly occurring proteins, such as cytochrome-c, may well be of considerable interest. Amino-acid sequences provide a large number of unit characters, which lend themselves to numerical analysis, and to the calculation of the probability that differences between organisms have arisen from different sequential substitutions. The amino-acid sequence of cytochrome-c has already been used to throw light on the phylogenetic history of prokaryote and eukaryote evolution (Dickerson 1980). Nucleotide sequences in DNA and RNA could be another source of new information of phylogeny in the future. For such purposes culture techniques are necessary to produce sufficient materials for analysis. *C. elegans* and several other nematodes can already be grown in mass culture, and so no doubt in time will other species.

9.2. Classifications

Traité de zoologie (Grassé 1965) gives a comprehensive classification of the nematodes, the work of several authors, with De Coninck responsible for the free-living nematodes. This classification can be seen as a logical development of the work of Filipjev (1918–21) and Chitwood (Chitwood and Chitwood 1950), who laid the foundations of later work on nematodes. Recently, Andrassy (1976) published a book entitled *Evolution as a basis for the systematization of nematodes* which rearranges the major groups of nematodes, but omits a discussion of the parasites of vertebrates. Andrassy's classification has met with much criticism (Coomans 1977) and it is too early to say how much of his sweeping reorganization will meet with general acceptance.

Quite apart from arguments as to the scientific basis for Andrassy's classification, his nomenclatorial changes have been received with dismay. For example, one of his subclasses Secernen*tia* corresponds closely to Chitwood's Secernen*tea*. Such a change is bound to lead to confusion in the future. The name Torquentia has been substituted for Chromadorida, and Penetrantia for Enoplida, without any very good reasons. There have also been changes in generic names which are likely to be extremely confusing. More recently, Lorenzen (1981*b*) has published a thoroughly revised classification of the Adenophorea, rigorously applying the precepts of 'cladistic' phylogenetics. Lorenzen's classification differs from Andrassy's, and that of many other authors, in that he has thoroughly reappraised many of the morphological characters used in taxonomy, and has systematically looked for and investigated new potentially useful morphological characters. Widening the range of characters available to the systematist is certainly worthwhile. A disadvantage of Lorenzen's classification is that it does not deal with the classification of the Dorylaimida, an extremely important group within the Adenophorea. A valuable and more comprehensive account of the development of nematode taxonomy has been published by Heip, Vincx, Smol, and Vranken (1982). The equivalents for the major divisions in important classifications are set out in Table 9.1.

I have felt it necessary to give an outline of the classification of free-living nematodes. However, because of the incomplete and often conflicting classifications at hand, the classification given in Table 9.2 is a compromise. I have used Lorenzen's (1981*b*) classification for the Adenophorea, excluding the Dorylaimida, taking it to the family level.

TABLE 9.1

Broadly equivalent taxonomic subdivisions of the Nematoda in different classifcations

Authors	Chitwood and Chitwood (1950)	Grassé (1965)	Andrassy (1976)	Lorenzen (1981b)
Class	APHASMIDIA	ADENOPHOREA	TORQUENTIA	ADENOPHOREA
Subclass				
Orders	Chromadorida	Chromadoria	Monhysterida	Chromadoria
		Monhysterida	Araeolaimida	Monhysterida
		Araeolaimida	Desmoscolecida	
		Desmoscolecida		
		Chromadorida	Chromadorida	Chromadorida
		Desmodorida		
Subclass		Enoplia	PENETRANTIA	Enoplia
Orders	Enoplida	Enoplida	Enoplida	Enoplida
				Trefusiida
		Dorylaimida	Dorylaimida	Dorylaimida
Class	PHASMIDIA	SECERNENTEA	SECERNENTIA	SECERNENTEA
Subclass				
Orders	Rhabditida	Rhabditida	Rhabditida	Rhabditida
		Tylenchida	Tylenchida	Tylenchida
		Strongylida	Strongylida	
		Ascaridida	Ascaridida	
	Spirurida	Spirurida	Spirurida	

TABLE 9.2

A classification of Nematoda

| | Class | ADENOPHOREA* |
| | Subclass | Chromadoria |

Order   MONHYSTERIDA

Superfamily	Monhysteroidea	Superfamily	Siphonolaimoidea
Family	*Monhysteridae*	Family	*Linhomoeidae*
	Xyalidae		*Siphonolaimidae*
	Sphaerolaimidae		Axonolaimoidea
	Desmoscolecoidea		*Axonolaimidae*
	Desmoscolecidae		
	Meyliidae		*Diplopeltidae*
			Coninckiidae
			Comesomatidae

Order   CHROMADORIDA

Suborder	Leptolaimina	Suborder	Chromadorina
Family	*Leptolaimidae*	Superfamily	Chromadoroidea
	Chronogasteridae		*Chromadoridae*
	Plectidae		*Ethmolaimidae*
	Teratocephalidae		*Neotonchidae*
	Peresianidae		*Achromadoridae*
	Aulolaimidae		*Cyatholaimidae*
	Haliplectidae		*Selachinematidae*
	Rhadinematidae		Desmodoroidea
	Tarvaiidae		*Desmodoridae*
	Aegialoalaimidae		*Epsilonematidae*
	Ceramonematidae		*Draconematidae*
	Tubolaimoididae		Microlaimoidea
	Paramicrolaimidae		*Microlaimidae*
	Ohridiidae		*Aponchiidae*
	Prismatolaimidae		*Monoposthiidae*
	Bastianiidae		
	Odontolaimidae		
	Rhabdolaimidae		

Order   ENOPLIDA   Subclass Enoplia

Suborder	Enoplina	Suborder	Tripyloidina
Division	Enoplacea	Family	*Tripyloididae*
Superfamily	Enoploidea		*Triodontolaimidae*
Family	*Enoplidae*		*Rhabdodemaniidae*
	Thoracostomopsidae		*Pandolaimidae*
	Anoplostomatidae		*Tobrilidae*
	Phanodermatidae		*Tripylidae*
	Anticomidae		
	Ironoidea		
	Leptosomatidae	Order	TREFUSIIDA
	Oxystominidae	Family	*Trefusiidae*
	Ironidae		*Onchulidae*
Division	Oncholaimacea		*Odontolaimidae*

Superfamily   Oncholaimoidea
   Family   *Oncholaimidae*
        Enchelidiidae

Lauratonematidae
Xenellidae

Order   DORYLAIMIDA†

Suborder   Mononchina
Superfamily   Mononchoidea
        Anatonchoidea
     Isolaimina
        Isolaimoidea
     Bathydontina
        Bathydontoidea
        Mononchuloidea

Suborder   Dorylaimina
Superfamily   Nygolaimoidea
        Dorylaimoidea
     Diphtherophorina
        Diphtherophoroidea
        Trichodoroidea
     Trichosyringina
        Mermithoidea‡
        Trichuroidea‡

Class   SECERNENTEA

Order   RHABDITIDA*

Superfamily   Rhabditoidae
   Family   *Rhabditidae*
        Bunonematidae
        Panagrolaimidae
        Mylolaimidae
        Cephalobidae

Superfamily   Diplogasteroidea
   Family   *Diplogasteridae*
        Pseudodiplogaster-
          oididae
      Rhabditida *incertae sedis*
        Brevibuccidae
        Chambersiellidae
        Cylindrocorpidae

Order   TYLENCHIDA**

Suborder   Tylenchina
Superfamily   Herterodoidea
        Hoplolaimoidea
        Tylenchoidea
        Anguinoidea
     Criconematina
        Criconematoidea
        Hemicycliophoroidea
        Tylenchuloidea
        Tylenchocriconema-
          toidea

Suborder   Hexatylina
Superfamily   Neotylenchoidea
        Allantonematoidea
        Sphaerularioidea‡
     Myenchina‡
        Myenchoidea

Order   APHELENCHIDA

Superfamily   Aphelenchoidea
        Aphelenchoidoidea

Order   STRONGYLIDA‡
       OXYURIDA‡
       ASCARIDIDA‡
       SPIRURIDA‡

---

* Classification taken from Lorenzen (1981*b*).
** Classification from Siddiqi (1980).
† Based on Coomans and Loof (1970).
‡ All parasites of other animals.

With the Dorylaimida, I have taken the classification only as far as superfamilies, basing this largely on Coomans and Loof (1970). The suborder Dorylaimina is extremely rich in families, genera, and species of terrestrial nematodes, which have been the subject of continual revision. I have been unable to find common ground between the arrangement of families by different authors. Important contributions have been made by Ferris (1971) who lists 31 families, Siddiqi (1969) who lists 17 families, and Andrassy (1976) who lists 28. Within the Secernentea, the Tylenchida has been revised a number of times in recent years, and the classification I have used, which takes the order as far as superfamilies, is that of Siddiqi (1980). This order contains the great majority of economically important plant pathogens as well as important parasites of insects. The discovery by Bedding that one nematode, *Deladenus siridicola*, had been classified in two separate families, the Allantonematidae for the parasitic form of this species and the Neotylenchidae for the free-living form of the same species, should be a caution to all nematologists when considering the importance of some of the morphological characters used to delineate families (Bedding 1968).

9.3. Identification

Permanent mounts of free-living nematodes should be made in anhydrous glycerol, with the cover slip supported to prevent undue flattening of the specimen, for example, with glass beads of diameter comparable to that of the specimen. The cover slips for specimens mounted in glycerol on microscope slides, can be ringed with White's cement or with Glyceel (= Zut) (Hopkins and Williams, Essex, UK), or some other substance which adheres to glass in the presence of glycerol. The basis for Glyceel is nitrocellulose dissolved in butyl acetate, toluol, and linseed oil. Marine nematodes, after fixation in buffered formalin in sea water, can be satisfactorily transferred to anhydrous glycerol from freshwater by allowing the water to evaporate from a solution of 5 per cent glycerol. Terrestrial and freshwater nematodes are more difficult to mount, tending to collapse. I have found De Grisse's (1969) method of transferring the nematodes via absolute ethanol to anhydrous glycerol satisfactory; one of many published methods. The nematodes, fixed in buffered formalin, are transferred to ethanol by standing at 37 °C in a watch-glass of water over 95 per cent ethanol in a vacuum desiccator, which has been briefly evacuated. A relatively small volume of 5 per cent glycerol in 95 per cent alcohol is added to the ethanol in the watch-glass and the solution allowed to evaporate in air at 37 °C to pure glycerol.

Nematodes can be handled in glycerol with an eyebrow hair glued to a wood stick. Many nematologists favour specially constructed aluminium microscope slides in which the nematode is sandwiched between two cover slips, facilitating examination of both sides of the nematode, though it is not difficult to temporarily remove the cover slip and rotate a specimen on a glass slide, if supported by glass beads and ringed with Glyceel.

Nematodes can be stained by cotton blue in glycerol, or other stains, but such stains do not aid taxonomic work. Interpreting structures of taxonomic importance, which may be semitransparent, flexible, and close to the resolving power of the light microscope, as well as often seen superimposed, because of the symmetry of the body, is inherently very difficult. More than anything else this deters zoologists from nematology. Scanning electron microscopy greatly facilitates the examination of external features of the head, cuticle, and the copulatory organs, but destroys the specimens. It is simpler than examining *en face* views of decapitated heads, a common practice. It can be used on specimens taken from glycerol mounts and rehydrated. Transmission electron microscopy is also invaluable in interpreting structures, but not as a routine taxonomic method, and it also destroys the specimen. It can replace the difficult art of cutting and examining sections of the head by light microscopy. Normarski interference microscopy is non-destructive and helps to resolve internal cytoplasmic structures as well as external cuticular structures.

Type collections are best preserved on slides in glycerol and have not deteriorated over many decades. The collections known to me are listed in Table 9.3.

Measurements are of great importance in the differentiation of species within genera. The relative proportions of a species may vary less than absolute measurements, and this has been the rationale for reporting measurements as ratios. The most widely used has been De Man's ratio (see Fig. 9.1), but there are serious objections to relying on ratios for taxonomic descriptions (Roggen and Asselberg 1971; Geraert 1968), because the ratios do not remain constant with differences in the size of

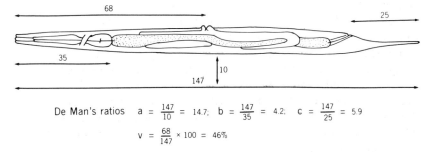

$$\text{De Man's ratios} \quad a = \frac{147}{10} = 14.7; \quad b = \frac{147}{35} = 4.2; \quad c = \frac{147}{25} = 5.9$$

$$v = \frac{68}{147} \times 100 = 46\%$$

Fɪɢ. 9.1. De Man's formula expressing measurements of a nematode as relative proportions.

TABLE 9.3

The location of some type collections which include free-living nematodes

The United States Department of Agriculture, Nematode Section, Beltsville, Maryland, USA

The University of California (Davis), Nematode Survey Collection, Davis, California, USA

The University of California (Riverside), Department of Nematology, Riverside, California, USA

Caenorhabditis Genetics Center, Division of Biological Sciences, 106 Tucker Hall, Columbia, Missouri 65211, USA

Rothamsted Experimental Station, Department of Nematodology, Harpenden, Herts, UK

Commonwealth Institute of Helminthology, St Albans, Herts, UK

British Antarctic Survey, Zoological Section, Marks Wood Experimental Station, Abbots Ripton, Hunts, UK

Landbouwhogeschool, Laboratorium voor Nematologie, Binnenhaven 15, Wageningen, Netherlands

Rijksuniversiteit, Instituut voor Dierkunde, Gent, Belgium

Laboratoire des Vers, Museum National d'Histoire naturelle, 43 rue Cuvier, 75005, Paris, France

Koninklijk Museum voor Midden Africa, Tervuren, Belgium

Nematodensammlung des Instituts fur Meeresforschung, Bremerhaven, am Handelshafen 12, D-285 Bremerhaven, Federal Republic of Germany

Government Forest Experiment Station, Laboratory of Forest Pathology Collection, Tokyo, Japan

Canadian National Collection of Nematodes, Entomology Research Institute, Ottawa, Canada

National Nematode Collection of New Zealand, Entomology Division, DSIR, Nelson, New Zealand

The Universitets Zoologiske Museum, Copenhagen, Denmark

The Zoological Institute, Academy of Sciences, Leningrad, USSR

British Museum (National History), Cromwell Road, London SW7 5BD, UK

National Nematode Collection, Division of Nematology, Indian Agricultural Research Institute, New Delhi-12, India.

individuals. Geraert suggests that the position of the vulva, relative to the distance from head to anus, and the position of the oesophageal bulb along the length of the oesophagus would be worth quoting. If ratios are to be given, then absolute measurements, with statistical measures of variation, should also be given. Fixatives alter measurements (Stone 1971), usually causing the body wall to shorten, often bending the oesophagus. It is worth noting that there may be differences in structure, quite apart from those of proportion, under different conditions of growth. For example, in *Acrobeloides* and *Cephalobus*, the structure of the head may show significant differences with differences in nutrition (Anderson 1968; Anderson and Hooper 1970).

It is impractical to attempt a comprehensive review of taxonomic publications dealing with lower taxonomic categories. An invaluable guide to taxonomic descriptions of many free-living nematodes is to be found in Gerlach and Riemann's (1973, 1974) two-volume work *The Bremerhaven checklist of aquatic nematodes, Parts I and II*. Moreover, a microfilm library of all the taxonomic papers listed can be purchased from these authors (Gerlach, personal communication: Instituts für Meeresforschung in Bremerhaven, Am Handelshafen 12, D2850, Bremerhaven—Federal Republic of Germany). Platt and Warwick are preparing an illustrated guide to British marine nematodes which from their unpublished manu-

TABLE 9.4
Some taxonomic publications of families of free-living nematodes

Taxon	Authors
Monhysterida (as recognized by Lorenzen (1981*b*))	
Axonolaimoidea	Wieser (1956)
Comesomatidae	Wieser (1954) (under Chromadoroidea); Jensen (1979*b*)
Desmoscolecidae	Freudenhammer (1975)
Linhomoeidae	Gerlach (1963*a*); Vitiello (1969)
Monhysteridae	Wieser (1956)
Xyalidae	Lorenzen (1977)
Chromadorida (as recognized by Lorenzen (1981*b*))	
Plectidae	Maggenti (1961)
Ceramonematidae	Haspeslagh (1979)
Chromadoridae	Wieser (1954); Inglis (1969)
Cyatholaimidae	Wieser (1954)
Desmodoridae	Gerlach (1953)
Epsilonematidae	Lorenzen (1973)
Draconematidae	Allen & Noffsinger (1978)
Microlaimidae	Jensen (1978) (includes Molgolaimidae)
Enoplida	
Enoploidea	Wieser (1953*a*)
Enoplidae	Inglis (1964)
Oncholaimidae	Rachor (1970)
Dorylaimida	
Dorylaimina	Siddiqi (1969); Ferris (1971); Loof and Coomans (1970)
Rhabditida	
Rhabditidae	Osche (1952); Dougherty (nomenclature only) (1955); Volk (1950); Sachs (1950); Massey (1974); Sudhaus (1974*a*, *b*, *c*, 1976, 1980*b*)
Diplogasteridae	Weingartner 1955
Cephalobidae	Thorne 1937

script is useful for Australian marine genera, so that it will probably be helpful in other parts of the world as well.

Many older taxonomic works are hard to obtain, though Chitwood and Chitwood's (1950) textbook has been reprinted and Filipjev's (1918–21) most important work has been translated into English and republished. Wieser (1959c) has published an important bibliography of marine nematodes. Cobb's (1920) *One hundred new nemas*, and monographs by Thorne and Swanger (1936) and Thorne (1939) on the Dorylaimida are important and very difficult to obtain. A paper by Loof and Coomans (1970), in an inaccessible publication, gives a valuable systematic treatment of taxonomically important oesophageal gland nuclei in the Dorylaimina. Schuurmans Stekhoven's (1935–6) contribution to the 'Fauna of the North and Baltic Seas' is also important.

I have received several major taxonomic works from Russian nematologists, e.g. *Free-living Nematodes of Baikala* (Tsalolichlin 1980) and *Free-living marine nematodes of the seas of the USSR* (Plantonova 1976). I have listed a number of taxonomic works on free-living nematodes in Table 9.4, but excluding plant-feeding nematodes. A booklet on *The biology and taxonomy of nematodes associated with bark beetles in the US* (Massey 1974) describes many free-living Rhabditida and Tylenchida. Two useful nomenclatorial reviews are *Check lists of the nematode families Dorylaimoidea, Rhabditoidea, Tylenchoidea and Aphelenchoidea* by Baker (1962) and a *Nomenclatorial compilation of plant and soil nematodes* by Trajan and Hopper (1974).

9.4. Chemotaxonomy

Gel electrophoresis of proteins, especially of enzymes, extracted from nematodes has proved a potentially useful method of elucidating the relationships between closely related species and distinguishing subspecific races. Polyacrylamide gel electrophoresis has been the most widely used technique. Serological methods of discriminating between the proteins extracted have also been used, but seem to have proved less effective than gel electrophoresis. In *Heterodera*, gel electrophoresis shows that different species give different patterns of protein bands, with lesser differences amongst the pathotypes of the species of these plant parasites (Trudgill and Carpenter 1971; Trudgill and Parrott 1972). The genera of *Meloidogyne*, *Ditylenchus*, and *Aphelenchus* differ in their protein bands (Dickson, Sasser, and Huisingh 1970), but more information can be obtained by comparing the bands associated with specific enzymes (following reactions with appropriate substrates) (Dickson, Huisingh, and Sasser 1971; Hussey and Krusberg 1971). Dehydrogenases (lactate, malate, glucose-6-phosphate, α-glycerophosphate), hydrolases (acid and alkaline phosphotase, esterases) show differences in their mobilities in the gels in different species of these genera and within populations of the same species. Isozymes, that is enzymes with the same substrate specificity but different mobility in the gels, have been found for malate dehydrogenase in species of *Ditylenchus* and *Meloidogyne* and for catalase in *D. triformis* (Dickson, Huisingh, and Sasser 1971; Hussey and Krusberg 1971). Esterase and dehydrogenase enzyme patterns distinguish *M. incognita* and *M. arenaria*, but α-glycerophosphate dehydrogenase and peroxidase isozymes may vary with the species of host plant (Hussey, Sasser, and Huisingh 1972).

Dalmasso and Berge (1978) analysed proteins and isozymes from single individuals of a number of species of *Meloidogyne* and correlated their findings with published work on chromosome complements (karyotypes). In this major study, tomato plants were used as host for 22 000 individuals from six species and 83 local populations. Nine strong protein bands were located by acrylamide gel electrophoresis, but not all were present in every species. Malate dehydrogenase, α-glycerophosphate dehydrogenase; esterases; and catalase were studied. Catalase gave a single band. There were two different esterases, differing in substrate specificity, which were interpreted as the expression of two separate gene loci in *M. hapla*, *M. arenarea*, *M. javanica*, and *M. incognita*. Each could give rise to two isozymes, i.e. enzymes with differing electrophoretic mobility, and these were interpreted as different alleles of a given locus. The alleles present differed in different populations and species. Similar interpretations were placed on the distribution of the other dehydrogenases in the various species and populations. The authors correlate these with the karyotyes and attempt to account for them according to the methods of reproduction.

References

Abdulkader, N. and Brun, J.L. (1978). *Rev. Nematol.* **1**, 27–37.
—— —— (1980). *Rev. Nematol.* **3**, 11–19.
Abirached, M. (1974). Thesis, University of Lyons.
—— Brun, J.L. (1975). *Nematologica* **21**, 151–62.
Aboul-Eid, H.Z. (1969a). *Nematologica* **15**, 437–50.
—— (1969b). *Nematologica* **15**, 451–63.
Abrams, B.I. and Mitchell, M.J. (1978). *Nematologica* **24**, 456–62.
Ahmad, I. and Jairajpuri, M. (1980). *Nematologica* **26**, 139–48.
—— —— (1981a). *Rev. Nematol.* **4**, 145–9.
—— —— (1981b). *Rev. Nematol.* **4**, 151–6.
Albertson, D. and Thomson, J.N. (1976). *Phil. Trans. R. Soc.* **B275**, 299–325.
Alexander, R.N. (1979). *The invertebrates.* Cambridge University Press.
Allen, M.W. and Noffsinger, E.M. (1978). *A revision of the marine nematodes of the superfamily Draconematoidea Filipjev, 1918 (Nematoda: Draconematina).* University of California Press, Berkeley.
Alongi, D.M. and Tietjen, J.H. (1980). In *Marine benthic dynamics* (ed. K.R. Tenore and B.C. Coull) pp. 151–66. University of South Carolina Press.
Anderson, G.L. and Dusenbery, D.B. (1977). *J. Nematol.* **9**, 253–4.
Anderson, R.V. (1968). *Can. J. Zool.* **46**, 309–20.
—— Byers, J.R. (1975). *Can. J. Zool.* **53**, 1581–95.
—— Coleman, D.C. (1981). *Nematologica* **27**, 6–19.
—— Darling, H.M. (1962). *Phytopathology* **52**, 722.
—— Hooper, D.J. (1970). *Can. J. Zool.* **48**, 457–69.
—— Kirchner, T.B. (1982). In *Nematodes in soil ecosystems* (ed. D. W. Freckman) pp. 157–77. University of Texas Press, Austin.
—— Elliott, E.T., McClellan, J.F., Coleman, D.C., Cole, C.V., and Hunt, H.W. (1978). *Microbial Ecology* **4**, 361–71.
Andrássy, I. (1956). In *English translation of selected East European papers in nematology* (ed. B.M. Zuckerman, M.W. Brzeski, and K.H. Deubert), pp. 73–84. University of Massachusetts Press, (1967).
—— (1976). *Evolution as a basis for the systematization of nematodes.* Pitman, San Francisco. [Translated from Hungarian.]
Andrew, P.A. and Nicholas, W.L. (1976). *Nematologica* **22**, 451–61.
Apin, P. and Kilbertus, G. (1981). *Rev. Nematol.* **4**, 131–43.
Atkinson, H.J. (1973a). *J. exp. Biol.* **59**, 255–66.
—— (1973b). *J. exp. Biol.* **59**, 267–74.
—— (1975). *J. exp. Biol.* **62**, 1–9.
—— (1977). *J. Zool., Lond.* **183**, 465–75.
—— (1980). In *Nematodes as biological models*, Vol. 2 (ed. B. M. Zuckerman), pp. 101–42. Academic Press, New York.
Atkinson, W.D. and Sharrocks, B. (1981). *J. anim. Ecol.* **50**, 461–71.
Aueron, F. and Rothstein, M. (1974). *Comp. Biochem. Physiol.* **49B**, 261–71.
Azmi, M.I. and Jairajpuri, M.S. (1976). *Nematologica* **22**, 277–83.
Baäth, E., Lohm, U., Lundgren, B., Rosswall, T., Söderström, B., Sohlenius, B., and Wirén, A. (1978). *Oikos* **31**, 153–63.
Bair, T.D. (1955). *J. Parasit.* **41**, 613–23.

Baker, A.D. (1962). *Check lists of the nematode superfamilies Dorylaimoidea, Rhabditoidea, Tylenchoidea and Aphelenchoidea*. E. J. Brill, Leiden.
Balan, J. and Gerber, N.N. (1972). *Nematologica* **18**, 163–73.
Balasubramanian, M. and Myers, R.F. (1971). *Exp. Parasit.* **29**, 330–6.
Baldwin, J.G. and Hirschmann, H. (1973). *J. Nematol.* **5**, 285–302.
—— —— (1975a). *J. Nematol.* **7**, 40.
—— —— (1975b). *J. Nematol.* **7**, 175–91.
—— —— (1976). *J. Nematol.* **8**, 1–17.
Banage, W.B. (1963). *J. Anim. Ecol.* **32**, 133–40.
—— (1964). *Proc. E. Afri. Acad.* **2**, 67–74.
—— (1965). *E. Afri. Agric. For. J.* **30**, 311–13.
—— (1966). *J. anim. Ecol.* **35**. 349–61.
Barrett, J. (1976a). In *The organisation of nematodes* (ed. N.A. Croll), pp. 11–70. Academic Press, New York.
—— (1976b) In *Biochemistry of parasites and hostparasite relationships*, (ed. H. van der Bossche), pp. 67–80. Elsevier/North Holland, Amsterdam.
—— Ward, C.W., and Fairbairn, D. (1971). *Comp. Biochem. Physiol.* **38B**, 279–84.
Barron, G.L. (1977). *The nematode-destroying fungi*. Canadian Biological Publications Ltd, Ontario.
Bassus, W. (1962). Wiss. Z. Humbolt.—Univ. Berlin Math.—*Naturwiss. Reike* **11**, 145–77.
B'Chir, M.M. (1979). *Nematologica* **25**, 22–31.
—— Dalmasso, A. (1979). *Rev. Nematol.* **2**, 249–57.
Beams, H.W. and Sekhon, S.S. (1972). *J. Ultrastruct. Res.* **38**, 511–27.
Bedding, R.A. (1968). *Nematologica* **14**, 515–25.
—— (1972). *Nematologica* **18**, 482–93.
Behme, R. and Pasternak, J. (1969). *Can. J. Genet. Cytol.* **11**, 993–1000.
Belar, K. (1923). *Biol. Zbl.* **43**, 413–18.
Bell, S.S. and Coull, B.C. (1978). *Oecologia (Berl.)* **35**, 141–8.
Bennet-Clark, H.C. (1976). In *The organisation of nematodes* (ed. N.A. Croll), pp. 313–42. Academic Press, New York.
Benton, A.W. and Myers, R.F. (1966). *Nematologica* **12**, 495–500.
Beguet, B. (1972). *Exp. Geront.* **7**, 207–18.
—— (1978). *Rev. Nematol.* **1**, 39–45.
—— and Brun, J.L. (1972). *Exp. Geront.* **7**, 195–206.
Bielka, H., Schultz, I., and Böttger, M. (1968). *Biochim. biosphys. acta* **157**, 209–12.
Bird, A.F. (1966). *Nematologica* **12**, 359–61.
—— (1968). *J. Parasit.* **54**, 475–89.
—— (1971). *The structure of nematodes*. Academic Press, New York.
—— (1980). In *Nematodes as biological models*, Vol. 2, (ed. B. M. Zuckerman), pp. 213–36. Academic Press, New York.
—— Buttrose, M.S. (1974). *J. Ultrastruct. Res.* **48**, 177–89.
—— McClure, M.A. (1976). *Parasitology* **72**, 19–28.
—— Rogers, C.E. (1965). *Nematologica* **11**, 224–30.
—— Downton, W.J.S., and Hawker, J.S. (1975). *Marcellia* **38**, 165–9.
Bird, G.W. (1970). *J. Nematol.* **2**, 404–9.
—— Mai, W.F. (1967). *Nematologica* **13**, 617–32.
Blackith, R.M. and Blackith, R.E. (1976). *Nematologica* **22**, 235–59.
Blake, C.D. (1961) *Nature, Lond.* 192, 144–5.
Boaden, P.J.S. and Platt, H.M. (1971). *Thalassia Jugoslavica* **7**, 1–12.

Bone, L.W. and Shorey, H.H. (1978). *J. Chem. Ecol.* **4**, 595–612.

Bonner, T., Menefee, M.G., and Etages, F.J. (1970). *Z. Zellforsch mikrosk Anat.* **104**, 193–204.

Boroditsky, J.M. and Samoiloff, M.R. (1973). *Can. J. Zool.* **51**, 483–92.

Bosher, J.E. and McKeen, W.E. (1954). *Proc. helminth. Soc. Wash.* **21**, 113–17.

Boucher, G. (1973). *Vie Milieu* **23**, 69–100.

Boveri, T. (1887). *Anat. Anz.* **2**, 688–95.

—— (1899). *Festschr. Kupffer*, pp. 383–90.

Brenner, S. (1974). *Genetics* **77**, 71–94.

Brenning, U. (1973). *Oikos* **15** (Suppl.), 98–104.

Bretschko, G. (1973). *Int. Ver. Theor. Angew. Limnol. Verh.* **18**, 1421–8.

Brockelman, C.R. and Jackson, G.J. (1978). *J. Parasit.* **64**, 803–9.

Brun, J.L. (1965). *Science* **150**, 1467.

—— (1966*a*). *Ann. Biol. anim. Biochim. Biophys.* **6**, 127–58.

—— (1966*b*). *Ann. Biol. anim. Biochim. Biophys.* **6**, 267–300.

—— (1966*c*). *Ann. Biol. anim. Biochim. Biophys.* **6**, 439–66.

—— (1967). *Nematologica* **12**, 539–56.

—— (1972). *Proc. XIth Int. Symp. Nemat.*, p. 4. Reading.

—— Cayrol, J.C. (1970). *Nematologica* **16**, 523–31.

Bryant, C. (1975). *Adv. Parasit.* **13**, 35–69.

—— (1978). *Adv. Parasit.* **16**, 311–31.

—— Nicholas, W.L., and Jantunen, R. (1967). *Nematologica* **13**, 197–209.

Buecher, Jr., E.J. and Hansen, E.L. (1971). *J. Nematol.* **3**, 199–200.

—— —— Yarwood, E.A. (1966). *Proc. Soc. exp. Biol. Med.* **121**, 390–3.

—— —— —— (1970). *Nematologica* **16**, 403–9.

—— Perez-Mendez. G., and Hansen, E.L. (1969). *Proc. Soc. exp. Biol. Med.* **132**, 724–8.

Burr, A.H. and Burr. C. (1975). *J. Ultrastruct. Res.* **51**, 1–15.

—— Webster, J.M. (1971). *J. Ultrastruct. Res.* **36**, 621–32.

Byerly, L., Cassada, R.C., and Russell, R.L. (1976). *Dev. Biol.* **51**, 23–33.

Byers, J.R. and Anderson, R.V. (1973). *J. Nematol.* **5**, 28–37.

—— —— (1972). *Can. J. Zool.* **50**, 457–65.

Cadet, P. and Dion, M. (1973). *Nematologica* **19**, 117–18.

Calcoen, J. and Roggen, D.R. (1973). *Nematologica* **19**, 408–10.

Caldwell, J.R. (1981). *Oikos* **37**, 160–6.

Canning, E.U. (1962). *Arch. Parsitenk.* **105**, 455–62.

—— (1973). *Nematologica* **19**, 342–8.

Capstick, C.K. (1959). *J. anim. Ecol.* **28**, 189–210.

—— Twinn, D.C., and Waid, J.S. (1957). *Nematologica* **2**, 193–201.

Carroll, J.J. and Viglierchio, D.R. (1981). *J. Nematol.* **13**, 476–82.

Cassada, R.C. and Russell, R.L. (1975). *Dev. Biol.* **46**, 326–42.

Castille, J.M. and Krusberg, L.R. (1971). *J. Nematol.* **3**, 284–8.

Castro, C.E. and Thomason, I.J. (1973). *Nematologica* **19**, 100–8.

Caveness, F.E. and Panzer, J.D. (1960). *Proc. helminth. Soc. Wash.* **27**, 73–4.

Cayrol, J.C. (1970). *Ann. Zool. ecol. anim.* **2**, 327–37.

Chalfie and Sulston, J. (1981). *Devl. Biol.* **82**, 358–70.

Chang, S.L., Berg, G., Clarke, N.A., and Kabler, P.W. (1960). *Am. J. Trop. Med. Hyg.* **9**, 136–42

Chantanao, A. and Jensen, H.J. (1969). *J. Nematol.* **1**, 166–8.

Chaudhuri, N. (1964). Thesis, University of Illinois.

Cheah, K.S. (1976*a*). In *Biochemistry of parasites and hostparasite relationships*, (ed. H. van der Bossche), pp. 133–43. Elsevier/North Holland, Amsterdam.

—— (1976*b*). In *Biochemistry of parasites and hostparasite relationships* (ed. H. van der Bossche), pp. 145–50. Elsevier/North Holland, Amsterdam.

Chen, T.A. and Wen, G.Y. (1972). *J. Nematol.* **4**, 155–61.

Cheng, R. and Samoiloff, M.R. (1971). *Can. J. Zool.* **49**, 1443–8.

Chernov, Ju. I., Striganova, R.R., and Anajeva, S.T. (1977). *Oikos* **29**, 175–9.

Chia, F. and Warwick, R.M. (1969). *Nature, Lond.* **224**, 720–1.

Chin, D.A. (1976). *Nematologica* **22**, 113–15.

—— Taylor, D.P. (1969*a*), *J. Nematol.* **1**, 313–17.

—— —— (1969*b*), *Nematologica* **15**, 525–9.

—— —— (1970). *Nematologica* **16**, 1–5.

Chitwood, B.G. and Chitwood, M.B. (1950). *An introduction to nematology.* Monumental Printing, Baltimore.

—— Murphy, D.G. (1964). *Trans. Am. microsc. Soc.* **83**, 311–29.

Chitwood, D.J. and Krusberg, L.R. (1980). *J. Nematol.* **12**, 217–18 (Abstract).

—— —— (1981*a*). *J. Nematol.* **13**, 105–11.

—— —— (1981*b*). *Comp. Biochem. Physiol.* **69B**, 115–20.

Chow, H.H. and Pasternak, J. (1969). *J. exp. Zool.* **170**, 77–84.

Chuang, S.H. (1962). *Nematologica* **7**, 317–30.

Clark, S.A., Shepherd, A.M., and Kempton, A. (1973). *Nematologica* **19**, 242–7.

Clark, W.C. (1960). *Nematologica* **5**, 178–83.

—— (1964). *N.Z.J. Agric. Res.* **7**, 441–3.

—— (1978). *J. Zool. Lond.* **184**, 245–54.

Clarke, A.J. and Perry, R.N. (1977). *Nematologica* **23**, 350–68.

Cobb, N.A. (1920). *One hundred new nemas. Contributions to a science of nematology* ix. Waverly Press, Baltimore.

Cole, C.V., Elliott, E.T., Hart, H.W., and Coleman, D.C. (1978). *Microbial Ecology* **4**, 381–7.

Cole, R.J. and Dutky, S.R. (1969). *J. Nematol.* **1**, 6.

—— Krusberg, L.R. (1967). *Exp. Parasit.* **21**, 232–9.

—— —— (1968). *Life Sci.* **7**, 713–24.

Coleman, D.C., Anderson, R.V., Cole, C.V., Elliott, E.T., Woods, L., and Campion, M.H. (1978). *Microbial Ecology* **4**, 373–80.

Colonna, W.J. and McFadden, B.A. (1975). *Archs. Biochem. Biophys.* **170**, 608–14.

Cooke, R.C. (1963). *Ann. appl. Biol.* **52**, 431–7.

—— (1968). *Phytopathology* **58**, 909–13.

—— Godfrey, B.E.S. (1964). *Trans. Br. mycol. Soc.* **47**, 61–74.

—— Pramer, D. (1968). Phytopathology **58**, 659–61.

Coomans, A. (1964). *Nematologica* (1963) **9**, 587–601.

—— (1965). *Nematologica* (1964) **10**, 601–22.

—— (1977). *Nematologica* **23**, 129–36.

—— (1978). *Annls Soc. r. zool. Belg.* **108**, 115–18.

—— (1979*a*). In *Root-knot nematodes (Meloidogyne species). Systematics, biology and control*, (ed. F. Lamberti and C.E. Taylor), pp. 1–19. Academic Press, New York.

—— (1979*b*). *Rev. Nematol.* **2**, 259–83.

—— De Coninck, L. (1963). *Nematologica* **9**, 85–96.

—— van der Heiden, A. (1971). *Z. Morph. Tiere* **70**, 103–18.

—— and Loof, P.A.A. (1970). *Nematologica* **16**, 180–96.

Cooper, Jr., A.F. and Van Gundy, S.D. (1970). *J. Nematol.* **2**, 305–15.

—— —— (1971). *J. Nematol.* **3**, 205–14.

Crofton, H.D. (1971). In *Plant parasitic nematodes* (ed. B.M. Zuckerman, W.F. Mai, and R.A. Rohde), Vol. I, pp. 83–113. Academic Press, New York.

Croll, N.A. (1966a). *J. Helminth.* **40**, 33–8.

—— (1966b). *Nematologica* **12**, 610–14.

—— (1967a). *Nematologica* **13**, 17–22.

—— (1967b). *Nematologica* **13**, 385–9.

—— (1970). *The behaviour of nematodes*. Edward Arnold, London.

—— (1975a). *J. Zool. Lond.* **176**, 159–76.

—— (1975b). *Can. J. Zool.* **53**, 894–903.

—— (1975c). *Adv. Parasit.* **13**, 71–122.

—— (1976a). In *The Organisation of Nematodes* (ed. N.A. Croll), pp. 341–61. Academic Press, New York.

—— (1976b). *Can. J. Zool.* **54**, 556–70.

—— (1977). *Ann. Rev. Phytopath.* **15**, 75–89.

—— Evans, A.A.F., and Smith, J.M. (1975). *Comp. Biochem. Physiol.* **51A**, 139–43.

—— Riding, I.L., and Smith, J.M. (1972). *Comp. Biochem. Physiol.* **42A**, 999–1009.

—— Smith, J.M. (1970). *Proc. helminth. Soc. Wash.* **37**, 1–5.

—— —— Zuckerman, B.M. (1977). *Exp. Aging Res.* **3**, 175–99.

—— de Soyza, K. (1980). *J. Nematol.* **12**, 132–5.

—— Viglierchio, D.R. (1969a). *J. Parasit.* **55**, 895–6.

—— —— (1969b). *Proc. helminth. Soc. Wash.* **36**, 1–9.

—— Wright, K.A. (1976). *Can. J. Zool.* **54**, 1466–80.

Crowe, J.H. and Madin, K.A. (1974). *Trans. Am. microsc. Soc.* **93**, 513–24.

—— —— (1975). *J. exp. Zool.* **193**, 335–43.

—— —— Loomis, S.H. (1977). *J. exp. Zool.* **201**, 57–64.

—— O'Dell, S.J., and Armstrong, D.A. (1979). *J. exp. Zool.* **207**, 431–8.

Cryan, W.S., Hansen, E., Martin, M., Sayre, F.W., and Yarwood, E.A. (1963). *Nematologica* **9**, 313–19.

Cuany, A. and Dalmasso, A. (1974). Abstracts *XII Internat. Symp. Nematology* 25, Granada.

Dalmasso, A. (1975). In *Nematode vectors of plant viruses*, (eds. F. Lamberti, C.E. Taylor, and J. W. Seinhorst). Plenum Press, London.

—— and Berge, J.B. (1978). *J. Nematol.* **10**, 323–32.

—— Younes, T. (1969). *Ann. Zool. Ecol. anim.* **1**, 265–79.

—— —— (1970). *Nematologica* **16**, 51–4.

Dasgupta, D.R. and Gunguly, A.K. (1975). *Nematologica* **21**, 370–84.

Debell, J.T. (1965). *Quart. Rev. Biol.* **40**, 233–51.

De Coninck, L. (1965). In *Traité de zoologie* (ed. P.P. Grassé), Vol. 4, pp. 1–27. Masson et Cie, Paris.

Decraemer, W. and Coomans, A. (1978). *Aust. J. mar. Freshwat. Res.* **29**, 497–508.

De Grisse, A.T. (1969). *Med. Fac. Landbouwwet.* Rijksuniv, Gent 34, 351–69.

—— (1977). *De ultrastruktuur van het zenuwstelsel in de kop van 22 soorten plantenparasitaire nematoden, behorende tot 19 genera (Nematoda: Tylenchida).* Rijksuniv, Gent. Belgium.

—— (1979). *Scanning electron microscopy* SEM Inc. AMF O'Hare, Illinois, USA.

—— Lagasse, A. (1969). *J. Microscopie* **8**, 677–80.

—— Lippens, P.L., and Coomans, A. (1974). *Nematologica* **20**, 88–95.

—— Natasasmita, S., and B'Chir, M. (1979). *Rev. Nematol.* **2**, 123–41.

Demeure, Y. (1978). *Rev. Nematol.* **1**, 13–19.
—— Freckman, D.W., and Van Gundy, S.D. (1979*a*). *J. Nematol.* **11**, 189–95.
—— —— —— (1979*b*). *Rev. Nematol.* **2**, 203–10.
Dennis, R.D.W. (1977*a*). *Int. J. Parasit.* **77**, 171–9.
—— (1977*b*). *Int. J. Parasit*, **77**, 181–8.
Deppe, U., Schierenberg, E., Cole, T., Krieg, C., Schmitt, D., Yoder, B., and von Ehrenstein, G. (1978). *Proc. Nat. Acad. Sci. USA* **75**, 376–80.
Deubert, K.H. and Rohde, R.A. (1971). In *Plant parasitic nematodes* (ed. B.M. Zuckerman, W.F. Mai, and R.A. Rohde) Vol. 2, pp. 73–90. Academic Press, New York.
—— Zuckerman, B.M. (1968). *Nematologica* **14**, 453–5.
Dickerson, R.E. (1980). *Scient. Am.* **242**, 99–110.
Dickson, D.W., Huisingh, D., and Sasser, J.N. (1971). *J. Nematol.* **3**, 1–16.
—— Sasser, J.N., and Huisingh, D. (1970). *J. Nematol.* **2**, 286–93.
Dinet, A. and Vivier, M.H. (1977). *Cah. Biol. mar.* **18**, 85–97.
—— (1979). *Cah. Biol. mar.* **20**, 109–23.
Dion, M. and Brun, J.L. (1971). *Molec. gen. Genetics* **112**, 133–51.
Doncaster, C.C. (1962). *Nematologica* **8**, 313–20.
—— (1966). *Nematologica* **12**, 417–27.
—— (1976). *J. Zool. Lond.* **180**, 139–53.
—— Seymour, M.K. (1973). *Nematologica* **19**, 137–45.
—— —— (1975). *Nematologica* (1974) **20**, 297–307.
Dougherty, E.C. (1953). *J. Parasit.* **39**, (Suppl.) **39**, 32.
—— (1955). *J. Helminth.* **29**, 105–52.
—— (1959). *Ann. NY Acad. Sci.* **77**, 27–54.
—— Calhoun, G.H. (1948). *Proc. helminth. Soc. Wash.* **15**, 55–68.
—— Hansen, E.L. (1956). *Proc. Soc. exp. Biol. Med.* **93**, 223–7.
—— —— (1957). *Anat. Rec.* **128**, 541–2.
—— —— Nicholas, W.L., Mollett, J.A., and Yarwood, E.A. (1959). *Ann. NY Acad. Sci.* **77**, 176–217.
Drechsler, C. (1941). *Biol. Rev.* **16**, 265–90.
Dropkin, V.H. (1963). *Nematologica* **9**, 444–54.
—— (1966). *Ann. NY Acad. Sci.* **139**, 39–52.
Duddington, C.L. (1951). *Trans. Br. mycol. Soc.* **34**, 322–31.
—— (1954). *Nature* **173**, 500–1.
—— (1955). *Bot. Rev.* **21**, 377–439.
Duggal, C.L. (1978*a*). *Nematologica* **24**, 213–21.
—— (1978*b*). *Nematologica* **24**, 257–68.
—— (1978*c*). *Nematologica* **24**, 269–76.
—— (1978*d*). *J. Zool. Lond.* **186**, 39–46.
Duncan, A., Schiemer, F., and Klekowski, R.Z. (1974). *Polskie Archwm Hydrobiol.* **21**, 249–58.
Dusenbery, D.B. (1973). *Proc. nat. Acad. Sci. USA.* **70**, 1349–52.
—— (1974). *J. exp. Zool.* **188**, 41–8.
—— (1975). *J. exp. Zool.* **193**, 413–18.
—— (1976*a*). *Comp. Biochem. Physiol.* **53C**, 1–2.
—— (1976*b*). *J. exp. Biol.* **198**, 343–52.
—— (1976*c*). *J. Nematol.* **8**, 352–5.
—— (1980). In *Nematodes as biological models* (ed. B.M. Zuckerman), Vol. 1, pp. 127–58. Academic Press, New York.
—— Sheridan, R.E., and Russell, R.L. (1975). *Genetics* **80**, 297–309.

Dutky, S.R., Robbins, W.E., and Thompson, J.V. (1967). *Nematologica* **13**, 140 (Abstract).
Ellenby, C. (1969*a*). *Nature, Lond.* **202**, 615–16.
—— (1969*b*) *Proc. R. Soc.* **B169**, 203–13.
—— Smith, L. (1966). *J. Helminth.* **40**, 323–30.
Ells, H.A. (1969). *Comp. Biochem. Physiol.* **29**, 689–701.
—— Read, C.P. (1952). *J. Parasit.* **38**, (Suppl.), 21.
—— —— (1961). *Biol. Bull. mar. Biol. Lab. Woods Hole* **120**, 326–36.
El-Sherif, M. and Mai, W.F. (1969). *J. Nematol.* **1**, 43–8.
Endo, B.Y. and Wergin, W.P. (1977). *J. Ultrastruct. Res.* **59**, 231–49.
Epstein, J. and Gershon, D. (1972). *Mech. Age Dev.* **1**, 257–64.
—— Castillo, J., Himmelhoch, S., and Zuckerman, B.M. (1971). *J. Nematol.* **3**, 69–78.
Erlanger, M. and Gershon, D. (1970). *Exp. Geront.* **5**, 13–19.
Evans, A.A.F. and Perry, R.N. (1976). In *The organisation of nematodes* (ed. N.A. Croll), pp. 282–424. Academic Press, New York.
Fatt, H.V. (1964). MA Thesis, University of California, Berkeley.
—— (1967). *Proc. Soc. exp. Biol. Med.* **124**, 897–903.
—— Dougherty, E.C. (1963). *Science* **141**, 266–7.
Faulkner, L.R. and Darling, H.M. (1961). *Phytopathology* **51**, 778–86.
Fenchel, T.M. and Riedl, R.J. (1970). *Mar. Biol.* **7**, 255–68.
Ferris, V.R. (1971). In *Plant parasitic nematodes* (ed. B.M. Zuckerman, W.F. Mai, and R.A. Rohde), Vol. 1, pp. 163–89. Academic Press, New York.
—— Goseco, C.G., and Ferris, J.M. (1976). *Science* **193**, 508–10.
Fielding, M.J. (1951) *Proc. helminth. Soc. Wash.* **18**, 110–12.
Filipjev, I.N. (1918–21). *Free-living marine nematodes of the Sevastopol area*: Issues 1 and 2. Russian Academy of Sciences, Petrograd, 1918; 1921. Translated 1968; 1970, I.P.S.T., Jerusalem.
—— (1929–30). *Arch. Hydrobiol.* **21**, 1–64.
Findlay, S.E.G. (1981). *Estuarine, Coastal Shelf Sci.* **12**, 471–84.
Fisher, K.D. (1968). *Nematologica* **14**, 7.
Foor, W.E. (1967). *J. Parasit.* **53**, 1245–61.
—— (1970). *Biology of Reproduction, Suppl.* **2** 177–202.
Franklin, M.T. and Hooper, D.J. (1962). *Nematologica* **8**, 136–42.
Freckman, D.W., Kaplan, D.T., and van Gundy, S.D. (1977). *J. Nematol,* **9**, 176–81.
Freudenhammer, I. (1975). Meteor Forsch-Ergebrisse, Reche D. No. 20, pp. 1–65, Berlin.
Friedman, P.A., Platzer, E.G., and Eby, J.E. (1977). *J. Nematol.* **9**, 197–203.
Gagarin, V.G. (1981). In *Evolution, Taxonomy, Morphology and Ecology of Free-living Nematodes*, (eds T. A. Platonova, and S.Ya. Tsaloliklin), pp 25–6. Academy of Sciences, USSR. (In Russian.)
Galtsova, V.V. (1981). In *Evolution, taxonomy, morphology and ecology of free-living nematodes*, (eds. T. A. Platonova and S.Ya. Tsaloliklin), Academy of Sciences, USSR. (In Russian.)
—— (1976). In *Nematodes and their role in the role in the Meiobenthos*, (ed. O.A. Skarlato) pp. 165–270. Nauka Publishing House, Leningrad. (In Russian.)
Garcea, R.L. Schachat, F., and Epstein, H.F. (1978). *Cell* **15**, 421–8.
Geraert, E. (1968). *Nematologica* **14**, 171–83.
—— (1972). *Ann. Soc. r. zool. Belg.* **102**, 171–98.
—— (1976). *Nematologica* **22**, 437–45.

—— (1978a). *Nematologica* **24**, 137–58.
—— (1978b). *Nematologica* **24**, 347–60.
—— (1979a). *Nematologica* **25**, 1–21.
—— (1979b). *Nematologica* **25**, 439–44.
—— (1980). *Ann. Soc. r. zool. Belg.* **110**, 73–86.
—— Grootaert, P., and Decraemer, W. (1980). *Nematologica* **26**, 255–71.
—— Sudhaus, W., and Grootaert, P. (1980). *Ann. Soc. r. zool. Belg.* **109**, 91–108.
Gerlach, S.A. (1953). *Z. Morph. Ökol. Tiere* **41**, 411–512.
—— (1955). *Physiologica comp. Oecol.* **4**, 55–73.
—— (1963a). *Zool. Jb. Syst.* **90**, 599–658.
—— (1963b). *Kieler Meeresforsch.* **19**, 68–103.
—— (1966). *Mitt. biol. BundAnst. Ld-u. Forstw.* **118**, 25–39.
—— (1967). 'Meteor' Forschungsergebrusse Reihe D, 2 Gebrüder Bomtraeger 1, Berlin 38.
—— (1971). *Oecologia (Berl.)* **6**, 176–90.
—— (1977). *Ophelia* **16**, 151–65.
—— Riemann, F. (1973). *Veröff. Inst. Meeresforsch. Bremerh.* (Suppl.) 4, 1–404.
—— —— (1974). *Veröff. Inst. Meeresforsch. Bremerh. (Suppl.)* **4**, 405–736.
—— Schrage, M. (1971). *Mar. Biol.* **9**, 274–80.
—— —— (1972). *Veröff. Inst. Meeresforsch. Bremerh.* **14**, 5–11.
—— —— Riemann, F. (1979). *Veröff. Inst. Meeresforsch. Bremerh.* **18**, 35–67.
Gershon, H. and Gershon, D. (1970). *Nature, Lond.* **227**, 1215–17.
Glaser, R.W. (1940). *Proc. Soc. exp. Biol. NY.* **48**, 512–14.
—— McCoy, E.G., and Girth, H.B. (1942). *J. Parasit.* **28**, 123–6.
Goffart, H. and Heiling, A. (1962). *Nematologica* **7**, 173–6.
Goldstein, P. and Triantaphyllou, A.C. (1978a). *Chromosoma* **68**, 91–100.
—— —— (1978b). *Chromosoma* **70**, 131–9.
—— —— (1979). *J. Cell Sci.* **40**, 171–9.
—— —— (1980). *J. Cell Sci.* **43**, 225–37.
—— —— (1981). *Chromosoma* **84**, 405–12.
Goodchild, C.G. and Irwin, G.H. (1971). *Trans. Am. microsc. Soc.* **90**, 231–7.
Goodey, J.B. (1963). *Nematologica* **9**, 468–70.
Goodrich, M., Hechler, H.C., and Taylor, D.P. (1968). *Nematologica* **14**, 25–36.
Grassé, P.P. (1965) (Ed.). *Traité de Zoologie*, Vol. 4. Némathelminthes. Masson et Cie, Paris.
Gray, J. (1968). *Animal locomotion*. Weidenfeld & Nicholson, London.
—— Lissmann, H.W. (1964). *J. exp.Biol.* **41**, 135–54.
Green, C.D. (1966). *Ann. appl. Biol.* **58**, 327–39.
—— (1977). *Behaviour* **61**, 130–46.
—— (1980). Helminth. Abs. **B49**, 81–93.
—— Plumb, S.C. (1970). *Nematologica* **16**, 39–46.
Greet, D.N. (1964). *Nature, Lond.* **204**, 96–7.
—— (1978). *Nematologica* **24**, 239–42.
—— Green, C.D., and Poulton, M.E. (1968). *Ann. appl. Biol.* **61**, 511–19.
Grootaert, P. (1976). *Biol. Jl Dodonaea* **44**, 191–202.
—— Coomans, A. (1980). *Nematologica* **26**, 406–31.
—— Jaques, A. (1979). *Nematologica* **25**, 203–14.
—— —— Small, R.W. (1977). *Med. Fac. Landbouwwet, Rijksuniv, Gent* **42**, 1559–63.
—— Lippens, P.L. (1974). *Z. Moroph Ökol. Tiere* **79**, 269–82.

—— —— Ali, S.S., and De Grisse, A.T. (1976a). *Med. Fac. Landbouwwet, Rijksuniv, Gent.* **41**, 995–1005.

—— —— —— —— (1976b). *Biol. Jl Dodonaea* **44**, 203–9.

—— Maertens, D. (1976). *Nematologica* **22**, 173–81.

—— Wyss, U. (1978). *Nematologica* **24**, 243–50.

—— —— (1979). *Nematologica* **25**, 163–75.

Guiran, G. de (1979). *Rev. Nematol.* **2**, 223–31.

Gupta, M.C., Singh, R.S., and Sitaramaiah, K. (1979). *Nematologica* **25**, 142–5.

Gysels, H. (1968). *Nematologica* **14**, 489–96.

Hammen, C.S. (1967). *Nematologica* **13**, 599–604.

Hansen, E.L. and Buecher, Jr., E.J. (1970). *J. Nematol.* **2**, 1–6.

—— —— and Evans, A.A.F. (1970). *Nematologica* **16**, 328–9.

—— —— and Yarwood, E.A. (1964). *Nematologica* **10**, 623–30.

—— —— —— (1973). *Nematologica* **19**, 113–16.

—— Cryan, W.S. (1966a). *Nematologica* **12**, 138–42.

—— —— (1966b). *Nematologica* **12**, 355–8.

—— Perez-Mendez, G., and Beucher, E.J. (1971). *Proc. Soc. exp. Biol. Med.* **137**, 1352–4.

—— Sayre, F.W., and Yarwood, E.A. (1961). *Experientia* **17**, 32–5.

Harris, H.E. and Epstein, H.F. (1977). *Cell* **10**, 709–19.

—— Tso, M.W., and Epstein, H.F. (1977). *Biochemisty* **16**, 859–64.

Harris, J.E. and Crofton, H.D. (1957). *J. exp. Biol.* **34**, 116–30.

Harrison, B.D., Robertson, W.M., and Taylor, C.E. (1974). *J. Nematol.* **6**, 155–64.

Haspeslagh, G. (1979). *Annls Soc. r. zool. Belg.* **108**, 65–74.

Hechler, H.C. (1962a). *Proc. helminth. Soc. Wash.* **29**, 19–27.

—— (1962b). *Proc. helminth. Soc. Wash.* **29**, 162–7.

—— (1963). *Proc. helminth. Soc. Wash.* **30**, 182–94.

—— (1967). *Proc. helminth. Soc. Wash.* **34**, 151–5.

—— (1968). *Proc. helminth. Soc. Wash.* **35**, 24–30.

—— (1970). *J. Nematol.* **2**, 125–30.

—— (1972). *J. Nematol.* **4**, 243–5.

—— Taylor, D.P. (1966a). *Proc. helminth. Soc. Wash.* **33**, 71–83.

—— —— (1966b). *Proc. helminth. Soc. Wash.* **33**, 90–6.

Hedgecock, E.M. and Russell, R.L. (1975). *Proc. nat. Acad. Sci. USA* **72**, 4061–5.

Herman, R.K. and Horvitz, H.R. (1980). In *Nematodes as biological models* (ed. B.M. Zuckerman), Vol. 1, pp. 227–61. Academic Press, New York.

Hertwig, P. (1920). *Arch. mikrosk. Anat. Entw–mech.* **94**, 303–37.

—— (1922). *Z. Wiss. Zool.* **119**, 539–58.

Hieb, W.F. and Rothstein, M. (1968). *Science* **160**, 778–80.

—— —— (1975). *Exp. Geront.* **10**, 145–53.

—— Stokstad, E.C.R., and Rothstein, M. (1970). *Science* **168**, 143–4.

Hiep, C. and Decraemer, W. (1974). *J. mar. biol. Ass. UK* **54**, 251–5.

—— Smol, N., and Absillis, V. (1978). *Mar. Biol.* **45**, 255–60.

Heip, C., Vincx, M., Smol, G., and Vranken, G. (1982). *Helminth. Abs.* **B31**, 1–310.

Hill, A.V. (1929). *Proc. R. Soc. Lond.* **B104**, 39–96.

Himmelhoch, S., Kisiel, M.J., and Zuckerman, B.M. (1977). *Exp. Parasit.* **41**, 118–23.

—— Orion, D., and Zuckerman, B.M. (1979). *J. Nematol.* **11**, 358–62.

—— and Zuckerman, B.M. (1978). *Exp. Parasit.* **45**, 208–14.

Hirsch, D. (1979). *Symp. Soc. Dev. Biol.* **37**, 149–65.
—— Oppenheim, D., and Klass, M. (1976). *Dev. Biol.* **49**, 200–19.
—— Vanderslice, R. (1976). *Devl Biol.* **49**, 220–35.
Hirschmann, H. (1962). *Proc. helminth. Soc. Wash.* **29**, 30–42.
Hirumi, H., Chen, T.A., Lee, K.J., and Maramorosch, K. (1968). *J. Ultrastruct. Res.* **24**, 434–53.
Hoeppli, R.J.C. (1926). *Trans. Am. microsc. Soc.* **15**, 234–5.
Hollis, J.P. (1957). *Phytopathology* **47**, 468–73.
Honda, H. (1925). *J. Morph. Physiol.* **40**, 191–233.
Hope, W.D. (1969). *Proc. helminth. Soc. Wash.* **36**, 10–29.
—— (1977). *Mikrofauna Meeresboden* **61**, 307–8.
Hopper, B.E., Fell, J.W., and Cefalu, R.C. (1973). *Mar. Biol.* **23**, 293–6.
—— Meyers, S.P. (1966). *Helgoländer wiss. Meeresunters* **13**, 444–9.
—— —— (1967*a*). *Bull. mar. Sci.* **17**, 47–517.
—— —— (1967*b*). *Mar. Biol.* **1**, 85–96.
—— —— Cefalu, R.C. (1970). *J. invert. Path.* **16**, 371–7.
Horn, D.H.S., Wilkie, J.S., and Thomson, J.A. (1974). *Experientia* **30**, 1109–10.
Huang, C.S. and Huang, S.P. (1974). *Nematologica* **20**, 9–18.
Humphreys, W.F. (1981). *J. anim. Ecol.* **50**, 543–61.
Hunt, R.S. and Poinar, G.O. (1971). *Nematologica* **17**, 321–2.
Hussey, R.S. and Krusberg, L.R. (1971). *J. Nematol.* **3**, 79–84.
—— Sasser, J.N., and Huisingh, D. (1972). *J. Nematol.* **4**, 183–9.
Hwang, S.W. (1970). *Nematologica* **16**, 305–8.
Hyman, L.H. (1951). *The invertebrates: Acanthocephala, Aschelminthes and Entoprocta*, Vol. 3. McGraw-Hill, New York.
Imbriani, J.L. and Mankau, R. (1977). *J. invert. Path.* **30**, 337–47.
Inglis, W.G. (1964). *Bull. Br. Mus. nat. Hist. (Zool.)* **11**, 265–376.
—— (1969). *Bull. Br. Mus. nat. Hist. (Zool.)* **17**, 152–204.
Jackson, G.J. (1973). *Expl Parasit.* **34**, 111–14.
Jairajpuri, M.S. and Azmi, M.I. (1977). *Nematologica* **23**, 202–12.
—— Khan, W.U. (1975). *Nematologica* **21**, 409–10.
Jansson, H.B. and Nordbring-Hertz, B. (1980). *Nematologica* **26**, 383–9.
Jantunen, R. (1964). *Nematologica* **10**, 419–24.
Jarman, M. (1970). *Parasitology* **6**, 475–89.
—— (1976). In *The organisation of nematodes*, (ed. N.A. Croll), pp. 293–312. Academic Press, New York.
Jatala, P., Jensen, H.J. and Russell, S.A. (1974). *J. Nematol.* **6**, 130–1.
Jennings, J.B. and Colam, J.B. (1970). *J. Zool. Lond.* **161**, 211–21.
—— Deutsch, A. (1975). *Comp. Biochem. Physiol.* **52A**, 611–14.
Jensen, H.J. (1967). *Pl. Dis. Reptr* **51**, 98–102.
—— Gilmour, C.M. (1958). *Pl. Dis. Reptr* **52**, 3–4.
—— Siemer, S.R. (1969). *J. Nematol.* **1**, 12–13.
—— Stevens, J.O. (1969). *J. Nematol.* **1**, 293–4.
Jensen, P. (1978). *Zool. Scripta* **7**, 159–73.
—— (1979*a*). *Estuar. Coast. Mar. Sci.* **9**, 797–800.
—— (1979*b*). *Zool. Scripta* **8**, 81–105.
—— (1981). *Mar. Ecol. Prog.* Ser. 4, pp. 203–6.
Johnson, C.D. and Stretton, A.O.W. (1980). In *Nematodes as biological models* (ed. B.M. Zuckerman), Vol. 2, pp. 159–95. Academic Press, New York.
Johnson, P.W. and Graham, W.G. (1976). *Can. J. Zool.* **54**, 96–100.
—— Van Gundy, S.D., and Thomson, W.W. (1970*a*). *J. Nematol.* **2**, 42–58.

────── ─── (1970*b*). *J. Nematol.* **2**, 59–79.
Johnson, S.R., Ferris, J.M., and Ferris, V.R. (1972). *J. Nematol.* **4**, 175–83.
────── ─── (1973). *J. Nematol.* **5**, 95–107.
────── ─── (1974). *J. Nematol.* **6**, 118–26.
Jones, F.G.W. (1977). *Nematologica* **23**, 123–5.
Jones, T.P. (1966). *Nematologica* **12**, 518–22.
Juario, J.V. (1975). *Veröff. Inst. Meeresforsch. Bremerh.* **15**, 283–337.
Kahan, D. (1969). *Int. Ver. Theor. Angew. Limnol. Verh.* **17**, 829–30.
Kaulenas, M.S. and Fairbairn, D. (1968). *Exp. Cell Res.* **52**, 233–51.
Kerry, B. (1980). *J. Nematol.* **12**, 253–9.
Kimble, J., Sulston, J., and White, J. (1979). In *Cell lineage, stem cells and determination* (ed. N. Le Douarin), pp. 59–68. INSERM Symposium No. 10, Elsevier/North Holland, Biomedical Press, Amsterdam.
Kimpinski, J. and Welch, H.E. (1971). *Nematologica* **17**, 308–18.
King, C.E. (1962). *Ecology* **43**, 515–23.
Kirchner, T.B., Anderson, R.V., and Ingham, R.E. (1980). *Ecology* **61**, 232–7.
Kisiel, M.J., Deubert, K.H., and Zuckerman, B.M. (1976). *Exp. Aging Res.* **2**, 37–44.
──── Himmelhoch, S., and Zuckerman, B.M. (1972). *Nematologica* **18**, 234–8.
──── Nelson, B., and Zuckerman, B.M. (1972). *Nematologica* **18**, 373–84.
──── Zuckerman, B.M. (1974). *Nematologica* **20**, 277–82.
Kito, K. (1982). *J. Faculty Sci., Hokkaido Univ. Ser. VI Zool.* **23**, 143–61.
Klekowski, R.Z. (1971). *Polskie Archwn Hydrobiol.* **18**, 93–114.
──── Wasilewska, L., and Paplinska, E. (1972). *Nematologica* **18**, 391–403.
────── ─── (1974). *Nematologica* **20**, 61–8.
Klingler, J. and Kunz, P. (1974). *Nematologica* **20**, 52–60.
Knobloch, N. and Bird, G.W. (1978). *J. Nematol.* **10**, 61–70.
Köhler, P. (1980). In *Industrial and clinical enzymology* (ed. Lj Vitale and V. Simeon), pp. 203–56. Pergamon Press, Oxford.
Krieg, C., Cole, T., Deppe, U., Schierenberg, E., Schmitt, D., Yoder, B., and von Ehrenstein, G. (1978). *Dev. Biol.* **65**, 193–215.
Krüger, E. (1913). *Z. Zool.* **105**, 87–124.
Krumbein, W.C. and Pettijohn, F.J. (1938). *Manual of sedimentary petrography.* Appleton-Centuary-Crofts, New York.
Krusberg, L.R. (1960). *Phytopathology* **50**, 9–22.
──── (1967). *Comp. Biochem. Physiol.* **21**, 83–90.
──── (1972). *Comp. Biochem. Physiol.* **41B**, 89–98.
Lamberti, F. (1969). *J. Nematol.* **1**, 94–5.
──── Taylor, C.E., and Seinhorst, J.W. (Eds.) (1975). *Nematode vectors of plant viruses.* Plenum Press, London.
Leake, P.A. and Jensen, J.H. (1970). *J. Nematol.* **2**, 351–5.
Lee, D.L. (1964). *Proc. helminth. Soc. Wash.* **31**, 285–8.
Lee, J.J., Tietjen, J.H., Stone, R.J., Muller, W.A., Rullman, J., and McEnery, M. (1970). *Helgoländer wiss. Meesunters* **20**, 136–56.
Lees, E. (1953). *J. Helminth.* **27**, 95–103.
Leushner, J.R.A. and Pasternak, J. (1975). *Dev. Biol.* **47**, 68–80.
Lippens, P.L. (1974*a*). *Z. Morph. Tiere* **78**, 181–92.
──── (1974*b*). *Z. Morph. Tiere* **79**, 283–94.
──── Coomans, A., De Grisse, A.T., and Lagasse, A. (1975). *Nematologica* (1974) **20**, 242–56.

References 225

Loof, P.A.A. and Coomans, A. (1970). *Proc. IX Int. Nem. Symp., Warsaw* (1967), pp. 79–160.
Loomis, S.H., Madin, K.A.C., and Crowe, J.H. (1980). *J. exp. Zool.* **211**, 311–20.
—— O'Dell, S.J., and Crowe, J.H. (1980). *J. exp. Zool.* **211**, 321–30.
López-Abella, D., Jiménez-Millán, F., and Garcia-Hidalgo, F. (1967). *Nematologica* **13**, 283–6.
Lopez, G., Riemann, F., and Schrage, M. (1979). *Mar. Biol.* **54**, 311–18.
Lorenzen, S. (1973). *Mikrofauna des Meeresbodens* **25**, 411–94.
—— (1977). *Veröff. Inst. Meeresforsch. Bremerh.* **16**, 197–261.
—— (1978). *Zool. Scripta* **17**, 175–8.
—— (1981a). *Veröff. Inst. Meeresforsch. Bremerh.* **19**, 89–114.
—— (1981b). *Veröff. Inst. Meeresforsch. Bremerh.*, Suppl. 7 (1981).
Lower, W.R., Hansen, E. and Yarwood, E.A. (1966). *J. exp. Zool.* **161**, 29–36.
Lu, N.C. (1980). *Fed. Proc.* **39**, 3.
—— Hieb, W.F., and Stokstad, E.L.R. (1974). *Proc. Soc. exp. Biol. Med.* **145**, 67–9.
—— —— —— (1976). *Proc. Soc. exp. Biol. Med.* **151**, 701–7.
—— Hugenberg, G. Jr., Briggs, G.M., and Stokstad, E.L.R. (1978). *Proc. Soc. exp. Biol. Med.* **158**, 187–91.
—— Newton, C., and Stokstad, E.L.R. (1977). *Nematologica* **23**, 57–61.
Lumsden, R.D. (1975). *Exp. Parasit.* **37**, 267–339.
Lyons, J.M., Keith, A.D., and Thomason, I.J. (1975). *J. Nematol.* **7**, 98–104.
McIntyre, A.D. (1969). *Biol. Rev.* **44**, 245–90.
—— and Murison, D.J. (1973). *J. Mar. Biol. Ass. UK* **53**, 93–118.
Mackenzie, Jr., J.M., Garcea, R.L. Jr., Zengel, J.M., and Epstein, H.F. (1978). *Cell* **15**, 751–2.
——, Schachat, F., and Epstein, H.F. (1978). *Cell* **15**, 413–9.
Madin, K.A.C. and Crowe, J.H. (1975). *J. exp. Biol.* **193**, 335–42.
Maggenti, A.R. (1961). *Proc. helminth. Soc. Wash.* **28**, 139–66.
—— (1964). *Proc. helminth. Soc. Wash.* **31**, 159–66.
—— (1970). *J. Nematol.* **2**, 7–15.
—— (1971). In *Plant parasitic nematodes* (ed. B.M. Zuckerman, W.F. Mai, and R.A. Rohde), Vol. 1, pp. 65–81. Academic Press, New York.
—— (1979). *J. Nematol.* **11**, 94–8.
Malakhov, V.V. (1981). In *Evolution, taxonomy, morphology and ecology of free-living nematodes* (eds. T.A. Platonova and S. Ya. Tsaloliklin) pp. 45–51. Academy of Sciences, USSR. (In Russian.)
Mankau, R. (1980) *J. Nematol.* **12**, 244–52.
—— and Imbriani, J.L. (1975). *Nematologica* **21**, 89–94.
—— —— Bell, A.H. (1976). *J. Nematol.* **8**, 179–81.
—— Prasad, N. (1977). *J. Nematol.* **9**, 40–5.
Mapes, C.J. (1965a). *Parasitology* **55**, 269–84.
—— (1965b). *Parasitology* **55**, 583–94.
Marchant, R. and Nicholas, W.L. (1974). *Oecologia (Berl.)* **16**, 237–52.
Marks, C.F., Thomason, I.J., and Castro, C.E.F. (1968). *Exp. Parasit.* **22**, 321–37.
Massey, C.L. (1974). United States Department of Agriculture Handbook No. 446, Washington.
Maupas, E. (1899). *Archs Zool. exp. gén.* Ser. 3, **7**, 563–628.
—— (1900). *Archs Zool. exp. gén.* Ser. 3, **8**, 463–624.
McLachlan, A. (1978). *Estuar. Coast. Mar. Sci.* **7**, 275–90.
McLaren, D.J. (1976a). *Adv. Parasit.* **14**, 195–265.

—— (1976*b*). In *The organization of nematodes* (ed. N.A. Croll), pp. 139–61. Academic Press, New York.

McNeil, S. and Lawton, J.H. (1970). *Nature, Lond.* **225**, 472–4.

Meagher, J.W. (1975). *Nematologica* **20**, 323–36.

Mercer, E.K. and Cairns, E.J. (1973). *J. Nematol.* **5**, 201–8.

Metcalf, H. (1903). *Trans. Am. microsc. Soc.* **24**, 89–102.

Meyers, S.P., Feder, W.A., and Tsue, K.M. (1963). *Science* **141**, 520–2.

—— Hopper, B.E. (1966). *Bull. mar. Sci.* **16**, 142–50.

—— —— (1967). *Helgoländer wiss. Meeresunters* **15**, 270–81.

—— —— Cefalu, R. (1970). *Mar. Biol.* **6**, 43–7.

Meyl, A.H. (1953*a*). *Z. Morph Ökol. Tiere* **42**, 67–116.

—— (1953*b*). *Z. Morph Ökol. Tiere* **42**, 159–208.

—— (1953*c*). *Z. Morph. Ökol. Tiere* **42**, 421–48.

——(1955). *Arch. Hydrobiol.* **50**, 568–614.

—— (1957). *Hydrobiologia* **53**, 520–6.

Miller, C.W. and Jenkins, W.R. (1964). *Nematologica* **10**, 480–8.

—— Roberts, R.N. (1964). *Phytopathology* **54**, 1177.

Milutina, I.A. (1981). In *Evolution, taxonomy, morphology and ecology of free-living nematodes.*, (ed. T.A. Platonova and S.Ya. Tsaloliklin) pp. 38–44. Academy of Sciences, USSR. (In Russian.)

Moore, H.B. (1931). *J. mar. biol. Ass. UK* **17**, 325–58.

Moore, P.G. (1971). *J. mar. biol. Ass. UK* **51**, 589–604.

Morgan, G.T. and McAllan, W. (1962). *Nematologica* **8**, 209–15.

Moss, W.W. and Webster, W.A. (1970). *J. Nematol.* **2**, 16–25.

Mounier, N. (1981). *Nematologica* **27**, 160–6.

Murad, J.L. (1970). *Proc. helminth. Soc. Wash.* **37**, 10–13.

Myers, R.F. (1965). *Nematologica* **11**, 441–8.

—— (1966). *Nematologica* **12**, 579–86.

—— (1967*a*). *Nematologica* **13**, 323.

—— (1967*b*). *Proc. helminth. Soc. Wash.* **34**, 251–5.

—— (1968) *Exp. Parasit.* **23**, 96–103.

—— (1971). *Exp. Parasit.* **30**, 174–80.

—— Buecher, E.J., and Hansen, E.L. (1971). *J. Nematol.* **3**, 197–8.

—— Krusberg, L.R. (1965). *Phytopathology* **55**, 429–37.

Narang, H.K. (1970). *Nematologica* **16**, 157–22.

—— (1972). *Parasitology* **64**, 253–68.

Natasasmita, S. and De Grisse, A. (1978). *Med. Fac. Landbouwwet. Rijksuniv. Gent* **43**, 779–94.

Nicholas, W.L. (1956). *Nematologica* **1**, 237–40.

—— (1962). *Nematologica* **8**, 99–109.

—— Dougherty, E.C., and Hansen, E.L. (1959). *Ann. NY Acad. Sci.* **77**, 218–36.

—— —— —— Holm-Hansen, G. and Moses, V. (1960). *J. exp. Biol.* **37**, 435–43.

—— Grassia, A., and Viswanathan, S. (1974). *Nematologica* **19**, 411–20.

—— Hansen, E.L., and Dougherty, E.C. (1962). *Nematologica* **8**, 129–35.

—— Jantunen, R. (1963). *Nematologica* **9**, 332–6.

—— —— (1964). *Nematologica* **10**, 409–18.

—— —— (1966). *Nematologica* **12**, 328–36.

—— McEntegart, M.G. (1957). *J. Helminth.* **31**, 135–44.

—— Stewart, A.C. (1978). *Nematologica* **24**, 45–50.

—— Viswanathan, S. (1975). *Nematologica* **21**, 385–400.

Nickle, W.R. and McIntosh, P. (1968). *Nematologica* **14**, 11–12 (Abstract).

References   227

Nigon, V. (1949). *Annls. Sci. nat. (Zool.)* **11**, 1–132.
—— (1951). *Bull. Biol. Fr. Belg.* **95**, 187–225.
—— Brun, J. (1955). *Chromosoma* **7**, 129–69.
—— Dougherty, E.C. (1949). *J. exp. Zool.* **112**, 485.
Nonnenmacher-Godet, J. and Dougherty, E.C. (1964). *J. Cell Biol.* **22**, 281–90.
Nordbring-Hertz, B. (1977). *Nematologica* **23**, 443–51.
—— Mattiasson, B. (1979). *Nature* **281**, 477–9.
Norton, D.C. (1978). *Ecology of plant-parasitic nematodes.* John Wiley & Sons, New York.
Ogunfowora, A.O. and Evans, A.F. (1977). *Nematologica* **23**, 137–46.
Ohthof, T.H.A. and Estey, R.H. (1963). *Nature, Lond.* **197**, 514–15.
Orcutt, D.M., Fox, J.A., and Jake, C.A. (1978). *J. Nematol.* 264–9.
Osche, G. (1952). *Zool. Jb. Syst.* **81**, 190–280.
Ott, J. and Schiemer, F. (1973). *Netherlands J. Sea Res.* **7**, 233–43.
Ouazana, R. (1974). *Proc. XII Symp. Internat. Nemat., Granada*, September 1974 (Abstract).
—— Brun, J. (1975). *C. r. Acad. Sci., Paris* **280D**, 1895–8.
Overgaard Nielsen, C. (1949). *Nat. jutl.* **2**, 1–131.
Pai, S. (1927). *Zool. Anz.* **74**, 257–70.
Paine, R.T. (1971). *Ann. Rev. Ecol. Syst.* **2**, 145–64.
Pasternak, J. and Leushner, J.R.A. (1975). *J. exp. Zool.* **194**, 519–28.
—— Samoiloff, M.R. (1970). *Comp. Biochem. Physiol.* **33**, 27–38.
—— —— (1972). *Can. J. Zool.* **50**, 147–51.
Patel, T.R. and McFadden, B.A. (1976). *Anal. Biochem.* **70**, 447–53.
—— —— (1977). *Archs. Biochem. Biophys.* **183**, 24–30.
—— —— (1978a). *Nematologica* **24**, 51–62.
—— —— (1978b). *Exp. Parasit.* **44**, 262–8.
Peach, M. (1950). *Trans. Br. mycol, Soc.* **33**, 148–53.
—— (1952). *Trans. Br. mycol. Soc.* **35**, 19–23.
Person, F. (1974). *C. r. Acad. Sci. Paris* **279D**, 1891–4.
—— Brun, J. (1974). *Ann. Zool. ecol. Anim.* **6**, 111–30.
Pertel, R. (1973). *Genetics* **74S**, 211 (Abstract).
Peters, B.G. (1928). *J. Helminth.* **6**, 1–38.
—— (1930). *J. Helminth.* **8**, 133–64.
Petriello, R.P. and Myers, R.F. (1971). *Exp. Parasit.* **29**, 423–32.
Pillai, J.K. and Taylor, D.P. (1967). *Nematologica* **13**, 529–40.
—— —— (1968). *Nematologica* **14**, 159–70.
Pinnock, C., Shane, B., and Stokstad, E.L.R. (1975). *Proc. Soc. exp. Biol. Med.* **148**, 710–13.
Platt, H.M. (1977a). *Estuar. Coast. Mar. Sci.* **5**, 685–93.
—— (1977b). *Cah. Biol. Mar.* **18**, 261–73.
—— (1978). *Annls Soc. r. Zool. Belg.* **108**, 93–101.
—— Warwick, R.M. (1980). In *The shore environment*, Vol. 2: Ecosystems (ed. J.H. Price, D.E.G. Irvine, and W.F. Farnham), pp. 729–59. Academic Press, New York.
Platonova, T.A. (1976). In *Nematodes and their role in the Meiobenthos* (ed. O.A. Skarlato) pp. 3–164. Nauka Publishing House in Leningrad. (In Russian.)
Platzer, E.G. and Eby, J.E. (1980). *J. Nematol.* **12**, 234 (Abstract).
Poinar, G.O., Jr. (1972). *Ann. Rev. Ent.* **17**, 103–22.
—— (1977). *Nematologica* **23**, 232–8.
Pollock, C. and Samoiloff, M.R. (1976). *Can. J. Zool.* **54**, 674–9.

Poole, R.W. (1974). *An introduction to quantitative ecology*. McGraw-Hill, Kogakusha, Tokyo.
Popham, J.D. and Webster, J.M. (1978). *Can. J. Zool.* **56**, 1556–63.
Por, F.D. and Masry, D. (1968). *Oikos* **19**, 388–91.
Potts, F.A. (1910). *Quart. J. microsc. Sci.* **55**, 433–84.
Pramer, D. and Stoll, N.R. (1959). *Science* **129**, 966–7.
Price, R. and Warwick, R.M. (1980). *Oecologia (Berl.)* **44**, 145–8.
Rachor, E. (1969). *Z. Morph. Ökol. Tiere* **66**, 87–166.
—— (1970). *Veröff. Inst. Meeresforsch. Bremerh.* **12**, 443–53.
Raski, D.J., Jones, N.O., and Roggen, D.R. (1969). *Proc. helminth. Soc. Wash.* **36**, 106–18.
Reise, K. and Ax, P. (1979). *Mar. Biol.* **54**, 225–37.
Reiss, U. and Rothstein, M. (1974). *Biochem. NY* **13**, 1796–800.
—— —— (1975). *J. Biol. Chem.* **250**, 826–30.
Reitz, M.S. Jr. and Sanadi, D.R. (1972). *Exp. Geront.* **7**, 119–29.
Rew, R.S. and Saz, H.J. (1974). *J. Cell Biol.* **63**, 125–35.
Riddle, D.L. (1978). *J. Nematol.* **10**, 1–16.
—— (1980). In *Nematodes as biological models* (ed. B.M. Zuckerman), Vol. 1, pp. 263–83. Academic Press, New York.
Riemann, F. (1972). *Z. Morph Ökol. Tiere* **72**, 46–76.
—— (1974). *Mikrofauna des Meeresbodens* **40**, 249–61.
—— (1975). *Int. Revue ges. Hydrobiol. Hydrogr.* **60**, 393–407.
—— (1976). *Zool. Jb. Syst.* **103**, 290–308.
—— (1977a). *Mikrofauna Meeresboden* **61**, 217–30.
—— (1977b). *Veröff. Inst. Meeresforsch. Bremerh.* **16**, 263–7.
—— Schrage, M. (1978). *Oecologia (Berl.)* **34**, 75–8.
Riffle, J.W. (1968). *Nematologica* **14**, 14 (Abstract).
Rodriguez, J.G., Wade, C.F., and Wells, C.N. (1962). *Ann. ent. Soc. Am.* **55**, 507–11.
Roggen, D.R. (1970a). *Nematologica* **16,** 532–6.
—— (1970b). *Nematologica* **16**, 605–6.
—— (1973). *Nematologica* **19**, 349–65.
—— (1975). In *Nematode vectors of plant viruses* (ed. F. Lamberti, C.E. Taylor, and J.W. Seinhorst), pp. 129–37. Plenum Press, London.
—— (1979). *Nematologica* **25**, 128–35.
—— Asselberg, R. (1971). *Nematologica* **17**, 187–9.
—— Raski, D.J., and Jones, N.O. (1966). *Science* **152**, 515–16.
—— —— —— (1967). *Nematologica* **13**, 1–16.
Rohde, P.A. (1960). *Proc. helminth. Soc. Wash.* **27**, 121–3.
Roman, J. and Hirschmann, H. (1969). *Proc. helminth. Soc. Wash.* **36**, 164–74.
—— Triantaphyllou, A.C. (1969). *J. Nematol.* **1**, 357–62.
Rosenbluth, J. (1965). *J. Cell Biol.* **25**, 495–515.
—— (1967). *J. Cell Biol.* **34**, 15–33.
—— (1969). *J. Cell Biol.* **42**, 817–25.
Rossner, J. and Perry, R.N. (1975). *Nematologica* **21**, 438–42.
Rothstein, M. (1963). *Comp. Biochem. Physiol.* **9**, 51–9.
—— (1965). *Comp. Biochem. Physiol.* **14**, 541–52.
—— (1968). *Comp. Biochem. Physiol.* **27**, 309–17.
—— (1974). *Comp. Biochem. Physiol.* **49B**, 669–78.
—— (1980). In *Nematodes as biological models* (ed. B.M. Zuckerman), Vol. 2, pp. 29–46. Academic Press, New York.

—— Cook, E. (1966). *Comp. Biochem. Physiol.* **17**, 683–92.
—— Coppens, M. (1978). *Comp. Biochem. Physiol.* **B61**, 99–104.
—— Gotz, P. (1968). *Archs Biochem. Biophys.* **126**, 131–40.
—— Mayoh, H. (1964). *Archs Biochem. Biophys.* **108**, 134–42.
—— —— (1965). *Comp. Biochem. Physiol.* **16**, 361–5.
—— —— (1966). *Comp. Biochem. Physiol.* **17**, 1181–8.
—— Nicholls, F., and Nicholls, P. (1970). *Int. J. Biochem.* **1**, 695–705.
—— Tomlinson, G.A. (1961). *Biochem. Biophys. Acta* **49**, 325–627.
—— —— (1962). *Biochem. Biophys. Acta* **63**, 471–80.
Rubin, H. and Trelease, R.N. (1976). *J. Cell Biol.* **70**, 374–83.
Rutherford, T.A. and Croll, N.A. (1979). *J. Nematol.* **11**, 232–40.
Samoiloff, M.R. (1973). *Nematologica* **19**, 15–18.
—— Balakanich, S., and Petrovich, M. (1974). *Nature, Lond.* **247**, 73–4.
—— McNicholl, P., Cheng, R., and Balakanich, S. (1973). *Exp. Parasit.* **33**, 253–61.
—— Pasternak, J. (1968). *Can. J. Zool.* **46**, 1019–22.
—— —— (1969). *Can. J. Zool.* **47**, 639–43.
—— Smith, A.C. (1971). *J. Nematol.* **3**, 299–300.
Santmyer, P.H. (1955). *Proc. helminth. Soc. Wash.* **22**, 22–5.
—— (1956). *Proc. helminth. Soc. Wash.* **23**, 30–6.
Sauer, M.R., Chapman, R.N. and Brzeski, M.W. (1979). *Nematologica* **25**, 482.
Sayre, F.W., Hansen, E.L., and Yarwood, E.A. (1963). *Exp. Parasit.* **13**, 98–107.
—— Lee, R.T., Sandman, R.P., and Perez-Mendez, G. (1967). *Archs Biochem. Biophys.* **118**, 58–72.
Sayre, R.M. (1969). *Trans Am. microsc. Soc.* **88**, 266–74.
—— (1973). *J. Nematol.* **4**, 258–64.
—— (1980). *J. Nematol.* **12**, 260–70.
—— Powers, E.M. (1966). *Nematologica* **12**, 619–29.
Schachat, F.H., Garcea, R.L., and Epstein, H.F. (1978). *Cell* **15**, 405–11.
—— Harris, H.E., and Epstein, H.F. (1977). *Cell* **10**, 721–8.
Schiemer, F. (1975). In *Ecological aspects of used-water treatment* (ed. C.R. Curds and H.A. Hawkes), pp. 269–87. Academic Press.
—— Duncan, A. (1974). *Oecologia (Berl.)* **15**, 121–6.
—— Loffler, H., and Dollfuss, H. (1969). *Int. Ver. Theor. Angew. Limnol. Verh.* **17**, 201–8.
Schuurmans Stekhoven Jr., J.H. (1935–6). In *Tierwelt der Nord-und Ostsee*, (ed. Grimpe and Wagler), pp. 1–173 Leipzig.
Scott, H.L. and Whittaker, F.H. (1970). *J. Nematol.* **2**, 193–203.
Sekiya, R. (1966). *Jap. J. Parasit.* **15**, 475–83.
Seymour, M.K. (1975*a*). *Nematologica* **20**, 255–60.
—— (1975*b*). *Nematologica* **21**, 117–28.
—— (1977). *Nematologica* **23**, 187–92.
—— Minter, B.A., and Doncaster, C.C. (1978). *Nematologica* **24**, 167–74.
—— Shepherd, A.M. (1974). *J. Zool. Lond.* **173**, 517–23.
Sharma, J., Hopper, B.E., and Webster, J.M. (1978). *Annls Soc. r. Zool. Belg.* **108**, 47–56.
Shepherd, A.M. (1981). *Nematologica* **27**, 122–5.
—— Clark, S.A. (1976*a*). *Nematologica* **22**, 1–9.
—— —— (1976*b*). *Nematologica* **22**, 332–42.
—— —— Dart, P.J. (1972). *Nematologica* **18**, 1–17.
—— —— Hooper, D.J. (1980). *Nematologica* **26**, 313–57.

Shepherd, A.M., Clark, S.A., and Kempton, A. (1973). *Nematologica* **19**, 31–4.
Sher, S.A. and Bell, A.H. (1975). *J. Nematol.* **7**, 69–83.
Siddiqi, M.R. (1969). *Nematologica* **15**, 81–100.
—— (1978). *Nematologica* **24**, 449–55.
—— (1980). *Helminth. Abstracts* Ser. B, **49**, 143–70.
Siddiqi, I.A. and Viglierchio, D.R. (1970). *J. Ultrastruct. Res.* **32**, 558–71.
—— —— (1977). *J. Nematol.* **9**, 56–82.
Sin, W.C. and Pasternak, J. (1970). *Chromosoma* **32**, 191–204.
Singh, R.H. and Sulston, J.E. (1978). *Nematologica* **24**, 63–71.
Sivapalan, P. and Jenkins, W.R. (1966). *Proc. helminth. Soc. Wash.* **33**, 149–57.
Skoolmun, P. and Gerlach, S.A. (1971). *Veröff. Inst. Meeresforsch. Bremerh.* **13**, 119–38.
Smith, L. (1965). *Comp. Biochem. Physiol.* **15**, 89–93.
Small, R.W. and Evans, A.A.F. (1981). *Rev. Nematol.* **4**, 271–70.
Smol, N., Hiep, C., and Govaert, M. (1980). *Annls Soc. r. Zool. Belg.* **110**, 87–103.
Sohlenius, B. (1968). *Pedobiol.* **8**, 340–4.
—— (1979). *Holarctic. Ecol.* **2**, 30–40.
—— (1980). *Oikos* **34**, 186–94.
Somers, J.A., Shorey, H.H., and Gaston, L.K. (1977). *J. Nematol.* **2**, 143–8.
Soyza, K. de (1973). *Proc. helminth. Soc. Wash.* **40**, 1–10.
Spurr, H.W. Jr. (1976). *J. Nematol.* **8**, 152–8.
Stauffer, H. (1925). *Zool. Jb. Syst.* **49**, 1–118.
Steiner, G. and Heinly, H. (1922). *J. Wash. Acad. Sci.* **12**, 367–96.
Stephenson, W.S. (1942). *Parasitology* **34**, 253–65.
—— (1945). *Parasitology* **36**, 158–64.
Stoll, N.R. (1959). *Ann. NY Acad. Sci.* **77**, 126–36.
Stone, A.R. (1971). *Nematologica* **17**, 167–71.
Storch, V. and Riemann, F. (1973). *Z. Morph. Tiere* **74**, 163–70.
Stringfellow, F. (1974). *Proc. helminth. Soc. Wash.* **41**, 4–10.
Stynes, B.A. (1980). *J. Nematol* **12**, 238 (Abstract).
Sudhaus, W. (1974*a*). *Faun. Ökol. Mitt.* **4**, 365–400.
—— (1974*b*). *Zool. Jb. Syst.* **101**, 173–212.
—— (1974*c*). *Zool.Jb. Syst.* **101**, 417–65.
—— (1976). *Nematologica* **22**, 49–61.
—— (1980*a*). *Nematologica* **25**, 75–82.
—— (1980*b*). *Zool. Jb. Syst.* **107**, 287–343.
Sukul, N.C. and Croll, N.A. (1978). *J. Nematol.* **10**, 314–17.
—— Das, P.K., and Ghosh, S.K. (1975). *Nematologica* **21**, 145–50.
Sulston, J.E. and Brenner, S. (1974). *Genetics* (Suppl.) **77**, 95–104.
Sulston, J., Dew, M., and Brenner, S. (1975). *J. comp. Neurol.* **163**, 215–26.
—— Horvitz, H.R. (1977). *Dev. Biol.* **56**, 110–56.
Swartz, F.J., Henry, M., and Floyd, A. (1967). *J. exp. Zool.* **164**, 297–308.
Swedmark, B. (1964). *Biol. Rev.* **39**, 1–42.
Tartar, T.A., Stack, J.P. and Zuckerman, B.M. (1977). *Nematologica* **23**, 267–9.
Taylor, C.E. and Robertson, W.M. (1971). *Nematologica* **17**, 303–7.
—— —— (1975). In *Nematode vectors of plant viruses* (ed. F. Lamberti, C.E. Taylor, and J.W. Seinhorst), pp. 253–75. Academic Press, London.
—— Thomas, P.R., Robertson, W.M., and Roberts, I.M. (1970). *Nematologica* **16**, 6–12.
Tchesunova, A.V. (1981). In *Evolution, taxonomy, morphology and ecology of*

free-living nematodes (ed. T. A. Platonova and S. Ya. Tsaloliklin), pp. 88–95. Academy of Sciences, USSR. (In Russian.)

Teal, J.M. (1962). *Ecology* **43**, 614–24.

—— Wieser, W. (1966). *Limnol. Oceanogr.* **11**, 217–22.

Thirungnam, M. (1976). *Exp. Parasit.* **40**, 149–57.

—— Myers, R.F. (1974). *Exp. Parasit.* **36**, 202–9.

Thomas, P.R. (1965). *Nematologica* **11**, 395–408.

Thorne, G. (1939). *Capita Zool.* **8**, 1–261.

—— Swanger, H.H. (1936). *Capita Zool.* **6**, 1–223.

Tietjen, J.H. (1967). *Trans. Am. microsc. Soc.* **86**, 304–6.

—— (1969). *Oecologia (Berl.)* **2**, 251–91.

—— (1971). *Deep-sea Res.* **18**, 941–57.

—— (1976). *Deep-sea Res.* **23**, 755–68.

—— (1977). *Mar. Biol.* **43**, 123–36.

—— (1980a). *Microbiology 1980. Am. Soc. Microbiol.*, pp. 335–8.

—— (1980b). *Estuar. Coast. Mar. Sci.* **10**, 61–73.

—— Lee, J.J. (1972). *Oecologia (Berl.)* **10**, 167–76.

—— —— (1973). *Oecologia (Berl.)* **12**, 303–14.

—— —— (1975). *Cah. Biol. mar.* **16**, 685–93.

—— —— (1977a). *Mikrofauna Meeresboden* **61**, 263–70.

—— —— (1977b). In *Ecology of marine benthos* (ed. B.C. Coull), pp. 21–35. University of South Carolina Press, Columbia.

—— —— Rullman, J., Greengart, A., and Trompeter, J. (1970). *Limnol. Oceanogr.* **15**, 535–43.

Townshend, J.L. and Blackith, R.E. (1975). *Nematologica* **21**, 19–25.

Tracey, M.V. (1958). *Nematologica* **3**, 179–83.

Tarjan, A.C. and Hopper, B.E. (1974). *Nomenclatural compilation of plant and soil nematodes.* Society of Nematologists.

Triantaphyllou, A.C. (1966). *J. Morph.* **118**, 403–14.

—— (1970). *J. Nematol.* **2**, 26–32.

—— (1971a). In *Plant parasitic nematodes* (ed. B.M. Zuckerman, W.F. Mai, and R.A. Rohde), Vol. 2, pp. 1–34. Academic Press, New York.

—— (1971b). *J. Nematol.* **3**, 183–8.

—— (1979). In *Root-knot nematodes (Meloidogyne species) systematics, biology and control* (ed. F. Lambert and C.E. Taylor), pp. 85–109. Academic Press, New York.

—— (1981). *J. Nematol.* **13**, 95–104.

—— Fisher, J.M. (1976). *J. Nematol.* **8**, 168–77.

—— Hirschmann, H. (1962). *Nematologica* **7**, 235–41.

—— —— (1966). *Nematologica* **12**, 437–42.

—— —— (1967). *Nematologica* **13**, 575–80.

Tribe, H.T. (1977). *Biol. Rev.* **52**, 477–507.

—— (1979). *Ann. appl. Biol.* **92**, 61–72.

Trudgill, D.L. and Carpenter, J.M. (1971). *Ann. appl. Biol.* **69**, 35–41.

—— Parrott, D.M. (1972). *Nematologica* **18**, 141–8.

Tsaloliklin, S.Ya. (1980). *Free-living nematodes of Baikal.* Nauka Publ. House, Siberian Div. Novosibirsk. (In Russian.)

Tzean, S.S. and Estey, R.H. (1981). *J. Nematol.* **13**, 160–7.

van der Heiden, A. (1975). *Nematologica* **20**, 419–36.

Vanderslice, R. and Hirsch, D. (1976). *Dev. Biol.* **49**, 236–49.

Vanfleteren, J.R. (1973). *Nematologica* **19**, 93–9.

—— (1975a). *Nematologica* **21**, 413–24.
—— (1975b). *Nematologica* **21**, 425–37.
—— (1976a). *Nematologica* **22**, 103–12.
—— (1976b). *Experientia* **32**, 1087–8.
—— (1978). *Ann. Rev. Phytopath,* **16**, 131–57.
—— (1980). In *Nematodes as biological models* (ed. B.M. Zuckerman), Vol. 2, pp. 47–79. Academic Press, New York.
—— Avau, H. (1977). *Experientia* **33**, 902–4.
—— Neirynck, K., and Huylebroeck, D. (1979). *Comp. Biochem. Physiol.* **62B**, 349–54.
Van Gundy, S.D. (1965). *Ann. Rev. Phytopath.* **3**, 43–68.
Viglierchio, D.R. (1974). *Trans Am. microsc. Soc.* **93**, 325–38.
—— Croll, N.A., and Gortz, J.H. (1969). *Nematologica* **15**, 15–21.
—— Siddiqi, I.A. (1974). *Trans Am. microsc. Soc.* **93**, 338–43.
Vitiello, P. (1969). *Tethys* **1**, 493–527.
—— (1974). *Ann. Inst. Oceanogr. Paris* **50**, 145–72.
Volk, J. (1950). *Zool. Jb. Syst.* **79**, 2–70.
Volz. P. (1951). *Zool. Jb. Syst.* **79**, 514–66.
von Ehrenstein, G. and Schierenberg, E. (1980) In *Nematodes as biological models* (ed. B.M. Zuckerman) Vol. I, pp. 1–71. Academic Press, New York.
—— —— Miwa, J. (1979). In *Cell lineages, stem cells and cell determination* (ed. N. Le Douarin), INSERM Symposium No. 10, pp. 49–58. Elsevier/North Holland Biomedical Press, Amsterdam.
Wallace, H.R. (1958a). *Ann. appl. Biol.* **46**, 74–85.
—— (1958b). *Ann. appl. Biol.* **46**, 86–94.
—— (1958c). *Ann. appl. Biol.* **46**, 662–8.
—— (1959a). *Ann. appl. Biol.* **47**, 131–9.
—— (1959b). *Ann. appl. Biol.* **47**, 350–60.
—— (1959c). *Ann. appl. Biol.* **47**, 366–70.
—— (1961). *Nematologica* **6**, 222–36.
—— (1970). *Nematologica* **16**, 249–57.
—— (1971). In *Plant parasitic nematodes* (ed. B.M. Zuckerman, W.F. Mai, and R.A. Rohde), Vol. 1, pp. 257–80. Academic Press, New York.
—— (1973). *Nematode ecology and plant disease.* E. Arnold, London.
—— Doncaster, C.C. (1964). *Parasitology* **54**, 313–26.
Wang, E.L.H. and Bergeson, G.B. (1978). *J. Nematol.* **10**, 367–8.
Ward, A.R. (1973). *Mar. Biol.* **22**, 53–66.
—— (1975). *Mar. Biol.* **30**, 217–25.
Ward, S. (1973). *Proc. nat. Acad. Sci. USA* **70**, 817–21.
—— (1976). In *The organisation of nematodes* (ed. N.A. Croll), pp. 365–82. Academic Press, New York.
—— Thomson, N., White, J.F. and Brenner, S. (1975). *J. comp. Neurol.* **160**, 313–38.
Ware, R.W., Clark, D., Crossland, K., and Russell, R.L. (1975). *J. comp. Neurol.* **162**, 71–110.
Warwick, R.M. (1971). *J. mar. biol. Ass. UK* **51**, 439–54.
—— (1977). In *Biology of benthic organisms* (ed. B.F. Keegan, P.O. Ceidigh, and P.J.S. Boaden), pp. 577–85. Pergamon Press, New York.
—— (1981). Oecologia (Berl.) **51**, 318–25.
—— Buchanan, J.B. (1970). *J. mar. biol. Ass. UK* **50**, 129–46.
—— —— (1971). *J. mar. biol. Ass. UK* **51**, 355–62.

—— Joint, I.R., and Radford, P.J. (1979). In *Ecological processes in coastal environments*, pp. 429–50. Blackwell, Oxford.

—— Price, R. (1979). *Estuar. Coast. Mar. Sci.* **9**, 257–71.

Wasilewska, L. (1970). *Ekologia polska* **18**, 429–42.

Waterston, R.H., Epstein, H.F. and Brenner, S. (1974). *J. molec. Biol.* **90**, 285–90.

Watson, B.D. (1965). *Quart. J. microsc. Sci.* **106**, 75–81.

Watson J.E., Pinnock, C.B., Stokstad, E.L.R., and Hieb, W.F. (1974), *Ann. Biochem. exp. Med.* **60**, 267–71.

Weingartner, I. (1955). *Zool. Jb. Syst.* **83**, 248–317.

Wen, G.Y. and Chen, T.A. (1976). *J. Nematol.* **8**, 69–74.

White, J.G., Southgate, E., Thomson, J.N., and Brenner, S. (1976). *Phil. Trans. R. Soc. Lond.* **B275**, 327–48.

Whitford, W.G., Freckman, D.W., Santos, P.F., Elkins, N.Z., and Parker, L.W. (1982). In *Nematodes in soil ecosystems*, (ed. D.W. Freckman) pp. 98–116. University of Texas Press, Austin.

Weiser, W. (1952). *J. Mar. biol. Ass. UK* **31**, 145–74.

—— (1953a). *Acta Univ. Lund* N.F. Adv. 2, **49**, 1–155.

—— (1953b). *Arkiv for Zoologi* Ser. 2, **4**, 439–84.

—— (1954). *Acta Univ. Lund* N.F. Adv. 2, **50**, 1–148.

—— (1956). *Acta Univ. Lund* N.F. Adv 2, **52**, 1–115.

—— (1959a). *Free-living nematodes and other small Invertebrates of Puget Sound beaches*. University of Washington Press, Seattle.

—— (1959b). *Limnol. Oceanogr.* **4**, 181–94.

—— (1959c). *Acta Univ. Lund* N.F. Adv 2, **55**, 1–111.

—— (1960). *Limnol. Oceanogr.* **5**, 121–37.

—— (1960). *Z. vergl. Physiol.* **43**, 29–36.

—— (1961). *Limnol. Oceanogr.* **6**, 262–70.

Willett, J.D. (1980). In *Nematodes as biological models*, (ed. B.M. Zuckerman), Vol. 1, pp. 197–225. Academic Press, New York.

—— Downey, W.L. (1973). *Comp. Biochem. Physiol.* **46B**, 139–42.

—— —— (1974). *Biochem. J.* **138**, 233–7.

—— Freckman, D.W., and Van Gundy, S.D. (1978). *J. Nematol.* **10**, 301–2.

—— Rahim, I. (1978a). *Comp. Biochem. Physiol.* **61B**, 243–6.

—— —— (1978b). *Comp. Biochem. Physiol.* **60B**, 403–5.

—— —— Bollinger, J. (1978). *Fed. Proc. Am. Soc. exp. Biol.* **37**, 1537.

Wilson, P.A.G. (1976). *J. Zool.* **179**, 135–51.

Winberg, G.G. (1971) (ed.) *Methods for the estimation of production of aquatic animals*. Academic Press, New York. (Translated A. Duncan).

Winkler, E.J. and Pramer, D. (1961). *Nature, Lond.* **192**, 472–3.

Wisse, E. and Daems, W.T. (1968). *J. Ultrastruct. Res.* **24**, 210–31.

Womersley, C. (1981). *Comp. Biochem. Physiol.* **70B**, 669–78.

—— Smith, L. (1981). *Comp. Biochem. Physiol.* **70B**, 579–86.

—— Thompson, S.N., and Smith, L. (1982). *J. Nematol.* **14**, 145–57.

Wood, F.H. (1973a). *Nematologica* (1974) **19**, 528–37.

—— (1973b). *NZ J. Bot.* **11**, 231–40.

—— (1973c). *NZ J. Agri. Res.* **16**, 373–80.

—— (1975). *Nematologica* **20**, 347–53.

Wright, D.J. (1975a). *Comp. Biochem. Physiol.* **52B**, 247–54.

—— (1975b). *Comp. Biochem. Physiol.* **52B**, 255–60.

—— Awan, F.A. (1978). *J. Zool. Lond.* **185**, 477–89.

—— Newall, D.R. (1976). In *The organisation of nematodes* (ed. N.A. Croll), pp. 163–210. Academic Press, New York.
—— —— (1980). In *Nematodes as biological models* (ed. B.M. Zuckerman), Vol. 2, pp. 143–64. Academic Press, New York.
Wright, K.A. (1965). *Can. J. Zool*, **43**, 689–700.
—— (1976). In *The organisation of nematodes* (ed. N.A. Croll), pp. 71–105. Academic Press, New York.
—— (1980). In *Nematodes as biological models*, Vol. 2 (ed. B.M. Zuckerman), Vol. 2, pp. 237–95. Academic Press, New York.
—— Hope, W.D. (1968). *Can. J. Zool* **46**, 1005–11.
Yarwood, E.A. and Hansen, E.L. (1968). *J. Parasit.* **54**, 133–6.
—— —— (1969). *J. Nematol.* **1**, 184–9.
Yeates, G.W. (1967). *NZ J. Sci.* **10**, 927–48.
—— (1968). *Pedobiologia* **8**, 173–207.
—— (1969). *Nematologica* **15**, 1–9.
—— (1970). *J. Nat. Hist.* **4**, 119–36.
—— (1971). *Pedobiologia* **11**, 173–9.
—— (1972*a*). *J. Nat. Hist.* **6**, 343–55.
—— (1972*b*). *Oikos* **23**, 178–9.
—— (1973*a*). *Oikos* **24**, 179–85.
—— (1973*b*). *NZ J. Sci.* **16**, 727–36.
—— (1973*c*). *NZ J. Sci.* **16**, 710–25.
—— (1974). *NZ J. Zool.* **1**, 171–7.
—— (1977*a*). *Soil Sci.* **123**, 415–22.
—— (1977*b*). *Oikos* **28**, 309.
—— (1978*a*). *NZ J. Agric. Res.* **21**, 321–30.
—— (1978*b*). *NZ J. Agric. Res.* **21**, 331–40.
—— (1979*a*). *Nematologica* **25**, 275–8.
—— (1979*b*). *J. Nematol.* **11**, 213–29.
—— (1979*c*). *NZ J. Zool.* **6**, 641–3.
—— (1980). *NZ J. Agric. Res.* **23**, 117–28.
Younes, T. (1969). *Ann. Zool. ecol. Anim.* **1**, 407–17.
Yuen, P.H. (1966). *Nematologica* **12**, 195–214.
—— (1967). *Can. J. Zool.* **45**, 1019–33.
—— (1968*a*). *Nematologica* **14**, 554–69.
—— (1968*b*). *Nematologica* **14**, 385–94.
—— (1971). *Nematologica* **17**, 1–12.
Zeelon, P., Gershon, H., and Gershon, D. (1973). *Biochemistry* **12**, 1743–50.
Zengel, J.M. and Epstein, H.F. (1980). In *Nematodes as biological models* (ed. B.M. Zuckerman), Vol. 1, pp. 73–126. Academic Press, New York.
Zimmermann, A. (1921). *Rev. Suisse Zool.* **28**, 357–80.
Zuckerman, B.M. (1974). *Helminth. Abs. Ser. A*, **43**, 225–39.
—— (Ed.) (1980). *Nematodes as biological models*, Vol. 2. Academic Press, New York.
—— Himmelhoch, S. (1980). In *Nematodes as biological models* (ed. B.M. Zuckerman), Vol. 2, pp. 3–28. Academic Press, New York.
—— —— Nelson, B., Epstein, J., and Kisiel, M. (1971). *Nematologica* **17**, 478–87.
—— —— Kisiel, M. (1973*a*). *Nematologica* **19**, 109–12.
—— —— —— (1973*b*). *Nematologica* **19**, 117.
—— Kahane, I., and Himmelhoch, S. (1979). *Exp. Parasit.* **47**, 419–24.

—— Kisiel, M., and Himmelhoch, S. (1973). *J. Cell Biol.* **58**, 476–80.

zur Strassen, O. (1896). *Arch. Entw. Mech. Org.* **3**, 27–105; 132–90.

AUTHOR INDEX

SUBJECT INDEX